Judging Judi

Judging Judi

Taking a Closer Look

Judith M.L. Day

This picture was taken late August, 1996, three weeks after I discontinued consuming the drug Tegretol. Feeling much better, I wanted to resume my career in Nursing Education, but was unsuccessful as I was considered, by the medical establishment, an unmedicated mentally ill patient with bipolar disorder, and it was too risky. It took seven stressful months to overturn that decision.

Copyright 2011 Judith M.L. Day
ISBN 978-1-257-80026-1

Acknowledgements

In my over 20 years of volunteer work in the world of Myalgic Encephalomyelitis / Chronic Fatigue Syndrome and Fibromyalgia (ME/CFS and FM), I have met and been in contact with thousands of people. Their stories always touch my heart as, without exception, their courage of having to deal with not only their illness but also the results of it, keep me focused.

Judi is such a person. As ill as she was, she chose to stand up for what she believed was unfair not only for herself but for others. She continued her fight for justice even when financially she was not able to do so. She wanted to spare others from having to experience the lack of support and medical assistance she was entitled to. When she could not afford professional legal help, she took it upon herself to become her own lawyer and taking her case through the many layers of legal appeals.

This latest challenge of writing a book is another way that Judi wants to help others avoid the pitfalls she had to endure and survive.

Lydia E. Neilson, MSM
Founder, Chief Executive Officer
National ME/FM Action Network
Canada

After reading this well documented account, there is no doubt in my mind that Judi did not receive justice. She presented the facts, ie, the various medical and legal reports, so that the reader knows that she was not selective of information just to support her side of the story. However, although Judi lost the legal battle, she is still a winner. Now that all the medical reports are in the "public domain", she has been able to tell her story about the supreme difficulty to obtain appropriate medical treatment for illnesses such as ME, FM and MCS. Treatment with potent psychotropic drugs obviously made her health deteriorate and caused physical damage. Although the good doctors may outnumber the poor ones, the latter should not be protected by the competent physicians. The Hippocratic Oath is very specific in "first do no harm".

Dr. Philipa D. Corning, Ph D, B SC, CD

As a person with ME/CFS and a former nurse, I feel that Judging Judi should be required reading for doctors and nurses. Judi's experience with uncaring professionals who prescribe toxic medications without regard to underlying conditions, would be a lesson in how not to practice. I'm amazed at Judi's strength, in her writing of such a painful time and in her ongoing activism.

Linda Whalen

Dedication

To my deceased parents, Clarence and Louise Sparkes, and my late sister Elizabeth Coyle, who had their lives shortened by errors in medical diagnosis and treatment.

To my late Uncle Max and Aunt Mary, who were like second parents to me. They both contributed immensely to positively shape my life.

To my brothers Jim, Ford and Max, for always making me feel welcome when I go back to visit our home. Despite our differences in the past, through love and understanding, let us keep being respectful of one another.

To my elders, Aunt June, Aunt Lillian, Uncle Charles, and Frances Stacey who continue to be a positive influence on me.

To my children, Karen and Brian, who will always be the most important living aspect of my life. When you are happy, I am happy. When you hurt, I hurt. When you smile, I smile. When you cry, I cry. I know I have caused you both a great deal of grief and worry this past fifteen years, possibly longer, struggling like I have with a chronic, invisible, misunderstood illness, while raising you as a single parent. As you mature and age, I am confident that love, forgiveness, remembrance, and understanding will be the threads that will hold and keep us together as a family.

To my three grandchildren, Liam, Roman, and Isabella. I cherish each moment we have together, and I am so happy that you are blessed with the good fortune of having two parents living together, nurturing and guiding you to adulthood. My prayer for you is that you and your parents will all stay together in love and harmony as long as you live.

When I die, it will not matter how many material things I have accumulated, or how much money I saved. These items will not last. The only thing that will matter is whether the role modeling, good influence, and lecturing that I did in order to instill good values and high morals in my children will be passed on to theirs. Whether I was successful or not depends upon them. I did my very best to teach the children well. I love you all.

Special Thanks

To my friend, James Flynn, who stood by me more than any other when I was going through the long court battle. He came with me every day he could, took notes, and was a solid rock of support and understanding. I will be forever grateful to him.

To Betty Ponder, Lorna Drew, Jannel MacIntyre, Dana Breckon, Randall McGregor, Greg McCarthy, Anne Macies, Glenna Hanley, and Edie snow who helped me proof read the chapters that gave me the encouragement to continue with this monumental task.

To Mona Bursey, Pat Harris, Pat Diamond, Betty Power, Frances Prowse, Brigitte Godden, Aarian Robinson, Betty Anderson, Mary Hayes, Nita Sparkes, Sandra Adey, Jim Flynn, Helen Fleming, Ford Sparkes, Karen Day, and Brian Day, who gave testimony or were there in the courtroom taking notes and being supportive, or just being understanding and kind, during my ordeal.

To Connie Petley, Judy Knee, Anne Gilles, Marva Smallwood, Jim Flynn, Jim Sparks, and Brigitte Godden, who provided me with housing, office space, and equipment during the many months I spent in St. John's, while my trials were taking place.

To Dr. John Martin, MD FRCP (C); Dr. Wayne Button, MD; Dr. Eleanor Stein, MD FRCP; Dr. Michael Nurse, MD. FRCP; Dr. Abayomi Oguinami, MD FRCP (C);

Dr. Leslie Philips, Pharmaceutical Chemist; and Ms. Sandra Gosse, Physiotherapist; whose expert testimonies were objective, truthful, and supportive.

To the late Dr. Allen Frecker, M.D. LM.C.C., F.R.C.P.(C), for his medical care, support and his honesty.

To Mr. Bob Buckingham, LLB, whose generosity and understanding were beyond what anyone could imagine.

To my cousins, Linda Crawford and Betty Randell, who have been caring and displaying an interest with unwavering love and support for me since I can remember.

To my niece, Melissa Ma Huynh, for designing the logo on the front cover of my book.

To my special friend Blanche Kelleher Thistle, for our never ending friendship since we were two years old.

Table of Contents

Acknowledgements v
Dedication vii
Special Thanks ix
Introduction xiii

Chapter 1.	*Admission to Waterford Psychiatric Hospital* .... *1* *May 15 – June 19, 1995*	
Chapter 2.	*Certification under Mental Health Act* *9* *May15, 1995*	
Chapter 3.	*Normal Mental Status except no insight* .... *17* *May 17, 1995*	
Chapter 4.	*Bipolar Disorder vs. Reactive Psychosis* .... *25* *May 19, 1995*	
Chapter 5.	*The Infamous Weekend* .... *31* *May 20, 1995*	
Chapter 6.	*Frustrations Mounting* .... *41* *May 22, 1995*	
Chapter 7.	*The Beginning of Forced Drug Therapy* .... *51* *May 24, 1995*	
Chapter 8.	*Recertification under Mental health Act* .... *63* *May 30, 1995*	
Chapter 9.	*My Mother's Death* .... *69* *June 1, 1995*	
Chapter 10.	*Discharge from Waterford Psychiatric Hospital* .... *79* *June 19, 1995*	
Chapter 11.	*The Summer of 1995* *85*	
Chapter 12.	*Admission to Health Science Centre* .... *91* *Nov. 27, 1995 - Dec. 12, 1995*	
Chapter 13.	*Struggling while taking Lithium* .... *107* *December 19, 1995 to August 1996*	
Chapter 14.	*A Six Month Battle to resume Employment* .... *113* *September 1996 to March 1997*	
Chapter 15.	*First Admission to St. Clare's Psychiatric Ward* .... *125* *Feb. 23, 1997-Mar. 3, 1997*	
Chapter 16.	*Finally Back to Work* .... *139* *April, 1997*	

Chapter 17.	***Second Admission to St. Clare's Psychiatric Ward*** *................................157* ***September 28, - October 1, 1997***	
Chapter 18.	***Back on Long Term Disability*** *...169*	
Chapter 19.	***Admission to St. Clare's Psychiatric Ward*** *...................................179* ***January 9-12, 1998***	
Chapter 20.	***Readmission to St. Clare's Psychiatric Ward*** *..............................191* ***January 14-19, 1998***	
Chapter 21.	***Finally a Diagnosis of CFS/FM/MCS*** *..203*	
Chapter 22.	***Taking my Life Back*** *...217*	
Chapter 23.	***In Pursuit of Justice*** *..225*	
Chapter 24.	***The Trials and Tribulation*** *...241*	
Chapter 25.	***Appeals and Closure*** *...249*	

Epilogue & Summary 265

About the Author 271

Introduction

The practice of psychiatry can lead to a number of ways to sift relentlessly and sometimes recklessly through incidents in people's lives in order to see them as candidates for psychiatric therapy. Subjective psychiatric examination and medicalization of human behavior, seemingly to justify the practice of psychiatric medicine, can lead to horrific impacts, particularly if a person is in a vulnerable state like when they are emotionally weak, confused, or perhaps even suffering from an undiagnosed physical illness.

Presently, the treatment by psychiatrists is mainly drug therapy, (chemicals) that substitute the normal constituents of the body; including substances both formed within it, or could be acquired through food. These sometimes dangerous drugs are often times prescribed too quickly; and for too long a period, resulting in many overlooked negative results.[1] According to Health Canada Statistics, "Psychotherapeutic agents are ranked as the second largest therapy class of drug sales in 2010 in Canada, with sixty one million prescriptions."

I believe that drug therapy is overused on oppressed people who, when removed from the stressors that caused the oppression, would be normal. These oppressed, vulnerable, emotionally wounded people are viewed as mentally ill, and thus given a psychiatric label. Once labeled, it is very difficult to lift or erase it, and even more challenging to get free from the dangerous medications that could cause neurotoxicity. I know because I became one of those oppressed patients, lost among the mentally ill and damaged by forced drug therapy.

The harder I tried to prove to the psychiatrists that I was not mentally ill, the more opposition I faced. The ongoing oppression resulted in incidents of anxiety due to extreme stress, while trying to be understood. These critical errant incidents were recorded carelessly by medical personnel, and the label of mental illness was branded on my forehead. I am not a quitter, and knew that I was not mentally ill, so I fought for my rights and freedom. I have finally won!

My battle with psychiatrists lasted almost three years, created intolerable stress in my life, and prevented me from living normally as I had done for fifty years. Unfortunately, once diagnosed with a mental illness, one is forever labeled. The stress that results from this labeling can cause an intense human response, which can drive a person to near insanity trying to prove that he or she indeed is not insane. Due to this misdiagnosis and maltreatment by psychiatry, I lost my career, credibility, financial security and nearly my life.

For many years, I was told by a psychiatrist that I was depressed. I knew that I did not feel gloom and doom for long periods of time. If I were sad, there were good reasons, feeling sick, physically weak and tired. That would make anyone feel depressed and anxious, especially when the lab results were normal even though I was feeling so ill. Physicians concluded that if no physical illness were found by medical testing to support my physical weakness, fatigue or sick feeling, I "must" be depressed. This theory that was applied to me has now proven to be seriously wrong.

A review of my medical history reveals many issues such as physical symptoms. These symptoms resulted in an admission to a psychiatric ward in 1983 for drug induced toxicity; a debilitating physical illness in 1993, when I could not work for four months; a confinement against my will in a mental institution in 1995; and six more brief admissions to psychiatric wards, due mainly to side effects of the prescribed drugs and acute stress. Finally, I realized something was not adding up with my medical treatment, therefore I went looking for answers.

---

[1] Canadian Centre of substance abuse, and Health Canada Statistics

Because I believed that something was seriously wrong, I studied my hospital records from the Waterford Psychiatric Hospital, St. John's, NL. Canada. I discovered mistakes that explained why I was diagnosed with bipolar disorder at the age of 50, and forced on dangerous drug therapy. I then studied all my medical records that were available in Newfoundland.

After reviewing all the records, I found out that many errors and omissions were made, not only during my first and only admission to the Waterford Psychiatric Hospital in 1995, but also during my admission to the Health Science Centre, St. John's five months later. For many years, I attempted to correct these mistakes that had caused me so much damage, and have the incorrect diagnosis removed from my medical files, but I was unsuccessful.

I began my quest for justice for being misdiagnosed, treated inappropriately, and injured, by first communicating with the Newfoundland and Labrador Medical Board in 1997, but without any positive results.

I insisted that something was physically wrong with me and that the psychotropic medications were making me worse, to the point where I was having serious neurological symptoms. Alzheimer's and/or Pic's disease were added to the differential diagnosis due to the moderate to severe memory impairment that I always felt was drug induced. Finally, I received the diagnosis of Fibromyalgia (fm) with symptoms also suggestive of chronic fatigue syndrome (cfs) and multiple chemical sensitivity (mcs). Major depression and fm/cfs/mcs have definite differences in symptoms when closely compared to each other.

Fm/cfs/mcs are conditions that have been recognized by the Center of Disease Control in Atlanta Georgia, for only the past twenty years. There are still many physicians who know very little about these chronic illnesses and the havoc they can play on a person's health. Some physicians still don't believe they even exist at all or are a manifestation of a psychiatric disorder. But according to Dr. Devin Starlanyl, M.D.,[1] "Fibromyalgia is not a mental illness, and doctors who imply such are bragging about their ignorance, to the detriment of patients everywhere."

With fm/cfs/mcs come many problems including sensitivities and adverse drug reactions, which I knew I had experienced many times. The pieces of the puzzle were never put together until I suffered a lifetime of symptoms and three years of misdiagnosis and maltreatment by psychiatry.

Although, for many years, I had battled fm/cfs/mcs and not major depression and then, I was diagnosed with manic depression, after I experienced psychosis from antidepressant drugs mixed with others stimulants known for their side effect of psychosis. The forced dangerous drug therapy for manic depression was the worst thing I could have been exposed to, as it made my symptoms worse, leading to more damage to my body and resulting in complete disability.

When I was reduced to a very ill state of health by psychotropic drugs, which according to electroencephalograms (EEGs), these drugs had reduced my seizure threshold and put me at risk for convulsions. After being on Epival, an anticonvulsant for four months, I finally realized something was terribly wrong with my treatment, as I felt I was slowly dying, so I escaped from the psychiatric trap.

In January 1998, I checked out of the hospital against medical advice, slowly discontinued the Epival, and then started on my long journey back to my life.

In the fall of 1998, I initiated a medical malpractice lawsuit during which time, over the next seven years, I spent fifty-five days in the courtroom, ending when the Supreme Court of Canada refused my application for Leave to Appeal.

I have included in this book excerpts from medical expert psychiatric reports by Dr. Eleanor Stein, Dr. Joseph Berger, Dr. Keith Pearce, Dr. David Addleman, and Dr. Teehan, about my medical treatment in the hands of psychiatry and their impressions of me.

There are many examples of psychiatrists' opinions who gave information that was never written on my medical records and who also never interviewed me. The misinformation that was presented to the judge during the trial negatively impacted his decision and subsequent outcome of the lawsuit.

---

[1] Letter to the editor, New Yorker in response to article "Hurting all Over, November 17, 2000 by Dr. D. Starlanyl

I hope you will read and judge for yourself if the conclusions presented in the expert reports and testimony were biased, defamatory, subjective, troubling, demoralizing, inaccurate, and inscribed on medical reports without evidence to substantiate their inclusion.

This book has been fifteen years in the making, and I want it to offer hope to the many people who are suffering from the symptoms of cfs/fm/mcs and possibly misdiagnosed. I pray it also gives them an understanding about the illness, the encouragement to embrace their differences, and the strength to treat themselves holistically, while taking advantage of the professional help from many resources that are available today. I wrote this book to help enable them to cope more effectively while dealing with these debilitating illnesses.

I hope also that physicians, nurses and all health care providers will read this book, so they too will realize that as long as mistakes continue to be swept under the rug, nobody learns, everybody loses, and some even lose their lives.

The drugs that caused my disability are Lithium, Haloperidol, Cogentin, Prozac and finally Epival, prescribed and sometimes forcefully given over three years. These drugs are potentially potent and dangerous. They could be damaging to the body and can possibly be deadly. Therefore patients must be diagnosed properly, monitored very closely for side effects, and the drugs discontinued before neurological, metabolic damage or death results.

The side effects I experienced included memory loss, inability to concentrate, imbalance, ringing in my ears, nausea and vomiting, numbness and tingling sensations, chronic fatigue, abnormal movements, muscle twitching and weakness, loss of strength in legs, shuffling instead of walking, shaking, sweating profusely, anxiety, palpitations, shortness of breath, retention of fluid, obesity and metabolic disorders.

While taking these drugs, medical tests showed abnormal electrocardiogram, EKG, abnormal electroencephalograms, EEGs, abnormal cognitive and motor tests, and abnormal thyroid functioning. Still I was being prescribed these psychotropic drugs and was considered mentally ill. People are led to believe that drugs are harmless and will help them, but at what cost to their health?

Statistics reveal that drug interactions and overdoses account for the deaths of thousands of Canadians annually and the amount being prescribed has risen considerably. (I often wonder how many of the prescriptions are not taken, but flushed down the toilet or thrown in the landfill, thus contaminating our environment.)

I kept my last prescription for Epival, 250 milligrams in the morning, and 500 milligrams at bedtime that was written by Dr. John Doucet, FRCP(c), on February 13, 1998, as a reminder of how I became disabled by psychotropic drugs and to never take another one again.

I am very lucky that I was a fighter and now a survivor. I could always speak up for myself, which I did but I was still ignored, misdiagnosed, and maltreated by physicians and nurses for three years, and permanently injured while trapped within psychiatry. I am also blessed that I lived to write this book and tell this incredible story.

If I suffered from bipolar disorder, I would not be drug free for the past thirteen years, mentally healthy, after going through the court system for almost ten years, surviving bankruptcy due to legal costs, and writing this book.

What about the injured people who are not as knowledgeable, courageous, and strong? Who speaks for them? Forced drug therapy, except in life and death, or last resort situations, is wrong and the people doing it should be held accountable when people are harmed by unmonitored drug interactions and overdoses.

The words in this book typed in *Italic* font are from notes that were compiled in a journal while I was a patient in the Waterford Hospital, a psychiatric institution in St. John's, Newfoundland, and detained there under certificate from May 15 to June 21, 1995; the admission to the Health Science Centre, from November 27 to December 10, 1995; and the St. Clare's Hospital twice in 1997 and again for two short stays in January, 1998.

Why, you may ask, did I write down my thoughts in a journal? I did that because my brother Ford encouraged me to do so, when he visited me at the Waterford Hospital and I wanted to have a record of what I was feeling and thinking as I was being stressed, confined against my will, and controlled unreasonably.

I knew at that time that I was misdiagnosed and suffered only due to drug interactions, and someday I would write this devastating experience in a book that perhaps may make a difference in psychiatric care in the future.

The words typed in **bold** font are actual documentation transcribed from the medical records that I obtained from the hospitals, after I went looking for answers regarding the misdiagnosis of bipolar disorder, instead of drug induced psychosis, which I thought was the problem from the beginning.

The words typed in "regular" font are added comments to explain situations and help clarify my thoughts during these difficult times of repeated hospitalizations.

I have bracketed twenty-four annotations from my journal that Dr. Keith Pearce, the medical expert for the Canadian Medical Protective Association, had written on his medical expert report, which he thought revealed hat I was indeed suffering from a major mood disorder.

I hope this book captures an interest in each reader so you can really understand what can actually happen to an individual in the hands of psychiatry, after you lose your autonomy and freedom and are treated like an uncommunicative animal. You be the judge!

It is so easy to take your rights and freedom for granted when you have never had them taken away. Once you have, and experience the torture, it leaves its mark on you forever. This is my story.

The practice of medicine is an art, not a trade; a calling, not a business /Osler

St. John's, Newfoundland with the famous Cabot Tower in the background, where this story took place.

Chapter 1

Admission to Waterford Psychiatric Hospital
May 15 – June 19, 1995

As a Registered Nurse, I have always felt that my opinions and advice were highly respected by my employers, managers, educators, as well as my colleagues. Even before I went to nursing school at all, I was a nurse's aid in Old Perlican Hospital, giving quality nursing care which included administering treatments, dispensing medications, delivering a child and preparing a body for its final resting place after death. Finally In the fall of 1963, I entered the Grace Hospital School of Nursing in St. John's, Newfoundland. I wanted to be a nurse since I was six years old.

Very early in my nursing career, in the 1970s, when I worked in Lachine, Quebec, I was told by a Director of Nurses that she noticed how well I handled everyday working situations, with maturity and tact. I realize self- praise is no praise, but these words were written on a reference as well as "excellent nurse on all counts."

In September 1979, after working for fifteen years on the mainland of Canada, mostly as an operating room nurse, in Labrador, Ontario, British Columbia, and Quebec, I returned to my home province, the island of Newfoundland, and was hired immediately by St. Clare's Hospital, St. John's as casual relief for the following summer of 1980.

The province of Newfoundland's name was officially changed December 6, 2001 and is now called the province of Newfoundland and Labrador.

In September, 1980, I took a leave of absence from St. Clare's Hospital and went to the Health Science Centre, St. John's to take a Post Basic course in Operating Room Technique and Management. This course was one of my main reasons for moving back to St. John's, as I needed to have the course, in order to advance in my career in Peri-operative Nursing. It was being offered there, and funded by the Unemployment Insurance Commission (later named Employment Insurance.)

After I completed my studies, in very good standing, I was hired permanently as Staff Nurse at the Health Science Centre. Within three months, I was Head Nurse in different specialties, and when the job of Clinical Educator was initiated in 1983, I applied for the position, and was the successful candidate.

I loved teaching and mentoring nurses, so this was my opportunity to create my own orientation program, which I knew was needed in operating rooms right across Canada. I also wanted to develop written nursing procedures to follow the policies that were being formulated as quality control programs were beginning to be a huge part of nursing practice and risk management.

I was delighted when I was chosen for the job. Even working fewer hours, I was making more money per hour. This position was considered educational, not management and that suited me better for what I wanted to accomplish within my career. I liked teaching more then managing as I love closely interacting with people.

I cherished every minute that I worked in that capacity, because there were always changes that I embraced, and new innovations to be added to my teaching curriculum and audits. I was in charge of the quality assurance program for the operating room as well. My job never became dull or boring. My supervisor would say "whatever you are doing, keep doing, because we are having good results."

Some say teachers are born. My mother was definitely born to be a teacher and perhaps so was I, but nursing was my passion. Now I had the best of both worlds. I could teach what I have always loved to do. Where could there be a better job for me? I felt there couldn't.

With input from every interested operating room nurse on staff, I formulated the Health Science Centre's written procedures according to the policies of the hospital. I followed the American Operating Room Nurse's Association, (AORN's), Recommended Standards of Practice, and they were given freely to the Operating Room Nurse's Association of Canada, ORNAC, which had its beginnings in 1983. These procedures, which consisted of two manuals, were used as guidelines while setting up the "Canadian Recommended Standards of Operating Room Nursing Practice". First edition in 1986, and the Canadian Recommended Technical Standards followed in 1988.

These two bodies of knowledge, including a chapter on competencies were combined by ORNAC in one single document and used as a primary resource for the Canadian Nurse's Association certification exam in peri-operative Nursing Practice in 1993,

My Supervisor was on the committees to do this volunteer work and I helped by writing the procedures that were used to set up the National Standards of Practice. Anne and I were very proud of what we had accomplished. By this time Anne was secretary of our national organization, and she gave me an honorary pin from ORNAC for my contribution to that association.

I worked in the responsible position of Clinical Educator, Operating Room, and was also in charge of the Quality Assurance Program until 1994, mostly part-time. The task however, would require me to work full-time hours for six weeks when new staff members were being orientated.

In 1994, with a bursary from the Health Science Centre, I took an educational leave of absence from my job to attend University of New Brunswick full time, as my ambition was to obtain a Masters Degree in counseling so that I could possibly continue with my nursing career at the Health Science Centre, with the plan to keep working well into my sixties.

At that point in my life, spring, 1995, I had worked in the nursing profession for over thirty years and had a very rewarding and successful career. Equipped with education, motivation and determination, I had been promoted from staff nurse to head nurse and then to clinical coordinator/ educator and now I wanted to be a counselor and a writer.

I took advantage of the university sponsored "free" tutoring and help with editing my written assignments and papers, and to further improve my writing skills, I also took an accredited course in creative writing.

While studying, I had a special interest in our Socialized Health Care System and compared ours to England's and United States'. For my creative writing course, I had researched and written a controversial paper called "Community Based Health Care, It is Plausible" and received very high marks for my effort, both from the Faculty of Nursing and Arts.

I had just returned from university to work for a few months, make some extra money, and then go back to complete my education; when one day in April, 1995, I saw a notice on the bulletin board of the Newfoundland Hospital Association's Writing Competition and discovered there was prize money involved.

I submitted my paper in the competition, as I had also been encouraged by my nursing professors to pursue a writing career and apply to the Government of Newfoundland to write for them. I thought this was an opportunity to get some exposure and make extra money in the meantime.

I did not win the competition: I came second, next to my new boss, who had written a paper on the subject of Mental Health. In the meantime, I believe I made enemies with the medical profession as I was advocating another more effective and less expensive method of Health Care services delivery.

I had believed strongly, which is still my belief, that the system we have in place in Canada is not Health Care: Our system is Illness Care and not enough consideration is given to the prevention of illness.

The Friday before I was certified to the Waterford Hospital, I was also asked to attend a Think Tank that brought together consumers and health care providers to discuss ways to save money in our Health Care System.

During those round table discussions, I presented my ideas that I had written in this paper on Health Care Reform and they were brought forward at the end of the day to the whole gathering.

Over the years, I had given many ideas to the hospital administration on how to save money. One time the hospital management gave out coffee mugs if our ideas were accepted. I had collected a dozen mugs during that competition.

I left the Think Tank gathering and drove to Sibley's Cove to spend the weekend with my parents and care for my mother. On the way, I stopped by Harbour Grace to see my brother Ford and his wife Joy, even though I had an upper respiratory infection, taking antibiotics and other drugs while feeling unwell.

I was pushing myself to the point of exhaustion on the weekend looking after Mom, as her home care workers only covered weekdays. It was Mother's Day; therefore I wanted to honor her request to help her attend church, as the wheelchair ramp had been newly built, which allowed her access.

I have written a poem about that Mother's Day. It tells it all!

Her last mother's Day, 1995

My mind goes back to Mother's Day,
Sixteen years ago, this year.
The only wish my mother had,
To go to church and say a prayer.

She needed help to get that wish,
as she was an invalid.
The wheel chair ramp was newly built,
Dad and I, our job we did.

She was so happy there in church.
It even made me shed a tear.
I told a very dear old friend,
"I know my mother's death is near".

"If upon my mother's grave,
I do not ever shed a tear,
It's just because they're all let out
For my mother, oh so dear."

My mother told me as a child,
She was so sad on Mother's Day.
While other children's rose was red.
On the alter, a white one she'd lay.

Her mother died when she was born.
She did not have a chance to give
A red rose to her loving mother,
Or talk to her about how to live.

My mother lost her first-born child.
It must be everyone's nightmare.
She never ever forgot the pain,
Even after fifty years.

> I gave my mom a red rose that year,
> When a darling child, so dear,
> Gave me a green one, so sincere.
> This rose told me that God was near.
>
> Whether the rose is red, white or green,
> made of plastic, silk or real,
> Given as a loving symbol,
> See how good it makes one feel.
>
> This Mother's Day let us all reflect
> On our lives and why we're here.
> Let's help each other through life's journey,
> And always hold our mothers dear.

Original poem composed Mother's Day, 1996, revised annually, and distributed.

I drove back home to St. John's Sunday afternoon, and on the way I visited my children's paternal grandmother to wish her Happy Mother's Day. I arrived home around 2100 and went straight to bed.

Monday, May 15, 1995

(I woke up Monday morning feeling miserable), like I had been run over by a five ton truck. I had been fighting bronchitis and sinusitis for weeks now, and had been prescribed antibiotics and antihistamines by Dr. Sharon Peters. I kept right on working as I usually did, going till I dropped as I have always had a good work ethic and held a responsible position as a Clinical Coordinator of the Operating Room at the Health Science Centre, St. John's, Newfoundland and Labrador, Canada, where I took my job very seriously.

I phoned my work-place to inform the manager that I was really sick and could not attend the seminar that was being held at the Radisson Hotel, sponsored by T.B. Cliffe. I was told that there was no other staff member slotted to go as they were short staffed and if I could not go, nobody would be attending.

Again I understood that it was important for someone to attend this seminar on "Cold Sterilization" and I being the Clinical Educator felt obligated, so (I decided to go despite how I felt).

Believe you me! I should have stayed in bed that day and perhaps what transpired for the next three years would never have happened.

I was physically ill and due to the fact that I was suffering from what was later diagnosed as fibromyalgia, chronic fatigue syndrome and multiple chemical sensitivities (fm/cfs/mcs), I had a psychotic reaction to the drugs I had taken.

(I did not eat breakfast, but took a Trinalin, to help clear my head, prescribed by Dr. Sharon Peters.) I was too ill to eat anything, so I drank a glass of orange juice, with that antihistamine and decongestant. I also took Prozac 10mgs, an anti-depressant that was prescribed by Dr. Subsash Jain, a psychiatrist, months before when he attributed the fatigue that I was experiencing to depression, even though I insisted that I did not have feelings of gloom and doom. I was just profoundly fatigued. As well, I took a couple of aspirins as every muscle in my body ached.

Dr. Jain was a psychiatrist whom I had seen on and off since 1983, when I was admitted to a general hospital psychiatric floor for ten days. At that time, I was suffering from the same symptoms. I was at home, in bed, sick with sinusitis and bronchitis, taking decongestants, antihistamines and Gravol. I began hallucinating, going in and out of consciousness, so my partner at the time, called Dr. Rogers an ophthalmologist, who roomed with him. He, in turn called Dr. Jain, a psychiatrist, who made a house-call. I remember him sitting at my bedside and asking me questions as I drifted in and out of consciousness. He prescribed an antipsychotic, hypnotic, Halcion and chlorpromazine on top of what I had already taken, I assume, to sedate me into a deeper sleep.

I did not settle down after I consumed the drugs that Dr. Jain had ordered for me, and in fact during the night I became worse and had a severe psychotic reaction with delusions and hallucinations.

I still have some memories of that event like a bad dream, where I was going back in time, and a lot of people were around me dressed in long robes. Everything was very confusing, but I don't remember getting out of bed.

I was home alone with my two children, Karen and Brian, who were fifteen and twelve at the time. According to what I was told, I must have scared them terribly. I apparently was trying to find Aunt Rachael in order to kill her. There is no living person named Aunt Rachael, but she is an ancestor who lived in the early nineteen hundreds. My Uncle Max would say she was his guardian angel, as he felt she saved his life when he was in an automobile accident in New York, where he lived as a young man.

When my partner returned in the morning, he and Karen took me to the Health Science Centre where I was admitted for ten days. I was heavily sedated and slept for the first few days and have no recollection of anything. My daughter visited me often, washed my hair, took care of me and I remember nothing about it. I had an electroencephalogram (EEG) (a test done to record the electrical potentials of the brain), which was normal, and other examinations, but have no memory of them.

All medications were withheld and within a few days, I recovered enough for interviews and more examinations, which I can still recall. After the medical consultations, family interviews, and tests were completed, my family and I were told that I had experienced a drug induced psychosis and if I were careful taking medications, especially Gravol, it probably would not ever happen to me again. Dr. Jain never mentioned to me that I should avoid[1] Halcion, however I believe now that a combination of it with Gravol caused my reaction.

My mother, who documented her daily activities, had written the above information in her journal in 1983 that confirmed that Dr. Jain had diagnosed me with a psychotic drug reaction at that time. The written diagnosis of "bipolar manic" never appeared on my health records until May 15, 1995.

My mother's journal was presented to the courts during my ten year lawsuit battle when I sued Dr. Karagianis and Dr. Craig for their mismanagement of the medical care that I had received from them in 1995 that escalated until January 1998, and left me physically disabled and professionally destroyed.

During that drug induced illness in 1983, I was discharged after ten days from hospital and over the next two weeks, I gradually tapered off the chlorpromazine 25 mg. that had been prescribed by Dr. Jain. I took my summer holidays to rest and regain my energy and then went back to work as usual. I was treated no differently by everyone. Nobody questioned the fact that I had spent a week in hospital in psychiatry, as it was understood that I had a psychotic drug reaction because I had recovered so quickly and completely. I was not labeled with any mental illness and my competence was never questioned, and I was taking absolutely no psychotropic drugs.

Then on May 15, 1995, unlike the admission in 1983, I vividly remember the car ride to the hospital (Health Science Centre.) from the Radisson Hotel. I felt I was rational, but very physically ill. I even remember the conversations with my colleagues. During my admission in the emergency department, (the doctors who examined me told me I was physically worn out and I needed rest.)

(The doctors also told me I should be observed for awhile). As I had nobody at home to care for me, they decided to admit me to the hospital. I definitely needed observation and supervision as I well remember what I was told had happened in June, 1983, when I was home alone with my children after I received medication while suffering from an upper respiratory infection.

During this illness in 1995, there were no beds available, in the psychiatric department, at the Health Science Centre and St. Clare's, so I voluntarily consented to be admitted to the Waterford Psychiatric Hospital as I thought that I was making the right decision. I was aware that a few months prior, all the hospitals had merged and were governed by one intercity board "St. John's Health Care

---

[1] compendium of Pharmaceutical and specialities (CPS) 2008, pg.1025

Corporation.", now "Eastern Health" and I was reassured that I would be transferred back to the Health Science Centre, as soon as there was a bed available.

As I had never personally attached any stigmata to mental illness, (I thought going to the Waterford hospital would be okay.) It would be the same as any hospital under that Board and I would be medically treated as I was at the Health Science Centre in 1983. This proved very wrong...I was harshly treated like an uncommunicative animal there.

I was transferred to the Waterford Hospital by taxi instead of ambulance, with no sedation, which was unusual if indeed the attending physicians had considered I was manic. I have a vivid memory of this taxi ride. I was accompanied by a female porter who told me she spent one year in nursing, but did not like it, so she quit after her first year.

During the fifteen minute taxi ride, this young girl treated me well. She portrayed all the qualities of a kind and caring person. I told her that it was a shame she did not complete her nursing education as she had a great deal to offer the profession.

Dr. Jamie Karagianis and Dr. David Pratt of The Waterford Hospital received the letter that came with me from the Health Science Centre with the written diagnosis of bipolar-manic, and they treated the condition of bipolar disorder (manic depression) instead of my actual symptoms. This mistaken diagnosis by the staff at the Health Science Centre, and passed along to the staff at the Waterford Hospital, nearly cost me my life.

I had no idea that the admitting diagnosis from the Health Science Centre was "bipolar-manic," and unaware, until I obtained my medical records December 1996, that this information was relayed by "letter" during my transfer to the Waterford Hospital. The actual letter was not included in my medical file, probably destroyed, so I have no idea who had written it.

When I reached the Waterford Hospital, I was escorted to the ward. All I wanted to do was find a bed and get into it, but I had to go through detailed interviews with Dr. James Karagianis and others. I felt so sick, but that made no difference to these doctors. It was as if my medical condition did not matter as I was constantly being interrogated and aggravated by them.

My son had just arrived home from Cuba that day and he and Sabrina, his partner at the time, (now his wife) came to the hospital immediately to see me and wanted to take me home. I however knew I was too sick to go home and that I required proper medical supervision and treatment.

Physician's Admission History:

Date, May 15, 1995 1630

This 49 year old female nurse from St. John's lives in her son's basement apartment and is separated from her second husband.

Chief complaint:

I've been working to hard, I need a rest, I'm afraid I am going insane.

History of Present illness:

Patient went to church yesterday (one day prior to arrival) with her mother and father. The day seemed stressful because patient is concerned about her mother's health and as well, it was Mother's Day and she did not hear from her son and daughter.

She woke up this morning (day of admission) and felt sick, so she called into work to try to get a replacement.

She ended up having to go to a meeting. Apparently it was an issue she did not agree with and she had an emotional outburst, where she became agitated and used offensive language.

She was brought to Health Science Centre for an assessment.

She was admitted to the Waterford Hospital because there were no beds available at Health Science Centre.

Medications:

Trinalin, antihistamine two by mouth for sinuses.

Allergies

Allergies, codeine,(nausea) and Gravol (agitation)

Psychiatric History:

She was admitted to Health Science Centre ten years ago with similar episode. Not being followed up since Dr. Jain left.

Past medical History:

Question

Circumstances prior to admission:

Patient is separated from second husband who has apparently been verbally abusive in the past. I did not think so. I did not say that.
She went to Fredericton and did seven courses to upgrade nursing.
Came back about four weeks ago and went back to work around three weeks prior to admission.
Her son was in Cuba on vacation. According to her son she has in the past shown episodes of hyperactivity and increase goal directed behavior.
She left home at sixteen, got her RN, has worked as RN in Vancouver, Montreal, St. John's. Married times two, now separated. Second husband verbally abusive. Son unaware of any abuse child/adult, no alcohol/drug abuse.

Family History:

Question mother depression. This snowballed into mother diagnosed with depression, which was untrue.
Active, energetic, "very intelligent"

MSE Mental status Exam:

Well groomed and dressed lady, looking younger than stated age. Her behavior changed drastically during interview.
Initially she was cooperative and behaved appropriately.
She became increasingly agitated and incoherent, banging the table and yelling "no, no, no." 'On my God Oh my God"
Affect labile with crying and laughing.
Speech initially normal in volume and then became loud.
Mood subjectively afraid, objectively first euthymic, later angry
Gross evidence of thought disorder with flight of ideas, worsening of associations and tangentiality.
Thought she could read other's mind
Initially orientated times three, later questionable.

Patient is a fairly intelligent woman, but seems to have no insight into her condition

Diagnosis:
1. **Bipolar disorder**
2. **Schizophrenia**

I consented to stay, signing Consent to Treatment forms that were part of my medical record, with no idea that I had been involuntarily certified immediately upon admission.

When I was informed by the nurse of being certified that first evening I was admitted, I became overwhelmed and very anxious about this situation as I could not understand why, and being a nurse, knew very well what "certification meant."

I was becoming more agitated and restless as I was physically ill and felt as if I were being badgered by Dr. Karagianis with questions that made no sense to me, especially when he brought me back into the room again and harassed me some more, after Brian stated that he wanted to take me home.

It was at this time that Dr. Karagianis kept asking me if I had any special powers and I said "no, no, no." I believe that Dr. Karagianis exaggerated my mental health status during those interviews and interrogated me unnecessarily, while I was very physically ill.

Patient's rule: it is not a matter of life and death, it's much more important than that. / anonymous

Waterford Psychiatric Hospital

Chapter 2

Certification under Mental Health Act
May 15, 1995

This is a replica of the Certification form that was signed by Dr. Karagianis and Dr. Pratt.

Schedule
Mental Health Act, 1971

The following facts and reasons for the above opinion of me the First physician, Dr. James Karagianis are as follows:

Forty nine year old female transferred from Health Science Centre with recent history of agitation and change of behavior.
She has history of episodes of being hyperactive with goal-directed behavior.
She has had an episode ten years ago. She can't provide much information. (I had informed them that I had a drug induced psychosis in 1983.)
Son said she is concerned about her mother dying soon. (She died two weeks later)
On examination she is alert, agitated with labile effect.
Repetitive speech, example Oh my God, Oh my God and no, no, no. She thought she could read my mind. (I said "I know what you are thinking")
Alternately crying/laughing
She pointed to a flower on her shirt that has special meaning. (It was given to me by a child in church the previous day.)
Pounding her fists on the table.
Obviously unable to look after herself.

The following facts and reasons for the above opinion of me the second physician, Dr. Douglas Pratt are as follows:

Patient is unable to explain why she is at the Waterford Hospital.
I had told them I was here because there were no beds available at the other hospitals. **Gross evidence of formal thought disorder with loosening of association, Tangential thought and incoherence. Vague reference to being in danger, but refuses to give details, acutely disorganized.**

The first evening the nurse told me "You are certified here and you must conform to treatment with drugs or be injected." I became really scared and anxious then. I knew that the main reason I had become so ill was the side effects of the drugs that I had consumed. I needed to detoxify, not be given more dangerous drugs to which I am sensitive. An ordinary aspirin had caused me to faint the first time I remember taking one.

I was told by a patient that I had hit her the first evening I was there. I don't remember doing that, and I was never told by the staff, the circumstances behind that incident. I am not a violent person so that puzzled me and I went looking for answers as I felt very badly about hitting somebody.

Later I learned from the patient, whom I had hit, that she thought I was going to die as I looked so ill. She was trying to place a crucifix around my neck, when I lashed back at her. She told me she felt sorry for me as she knew I was very sick

I then understood why I fought back. I must have been awakened, thought someone was trying to choke me and tried to protect myself. The patients could see I was very ill. Why didn't the doctors notice that as well?

The nurse's assessment, according to the form on my chart was outlining the correct information, "pale looking", "very weak and tired", coughing etc...but many doctors neither pay attention to what nurse's write on the patient's chart nor listen to them, so that information went unnoticed by Dr. Karagianis.

Nurse's Progress Notes

May 15, 1995

Admission to N3B, a 49 year old female under the services of Dr. Karagianis. Diagnosis of Bipolar-manic. She is allergic to Codeine. It is her first admission to the Waterford; however had previous admissions at Health Science Centre for same. (This was inaccurate information)

Having apparent difficulty concentrating, speech was appropriate and coherent, dress was appropriate and neat. She was orientated times three.

She has poor insight into her illness. She reported that she merely needs a rest because she is so tired. Does not believe she needs medication. She is very distressed because she is at the Waterford. She stated that she was only here due to the lack of beds in other hospitals. Oriented to the unit and placed on close observation.

The very first day of admission the diagnosis of bipolar disorder was established on my record along with the inaccurate information that I had had several admissions before at Health Science Centre.

Nursing Assessment Record,
May 15, 1995, 1330 hours

Reasons for Admission:

Bipolar & Manic

Patients Description of present Condition:

Tired, here for a rest. At the Waterford because there aren't any beds elsewhere. Felt sick and tired.

Past history pertinent to this admission:

First admission to Waterford however had previous admissions to HSC. (Inaccurate)

Non verbal actions:

Appeared to have difficulty concentrating, conversations contained (+++) sighs and oh my's. Good eye contact.

Appearance:

Neat, quiet lady, dress appropriate.

Activity:

Sitting quietly in chair, very little activity.

Comments:

Communication was coherent and made sense; however it appeared that she had difficulty concentrating on the topic. Thus was slow to answer any questions.

Appeared to shift, at one moment she was smiling and voice at normal tone and then she would switch to be quiet, confused and disorganized.

Orientated times three.

Memory:

She couldn't remember where her son lived.
Later she reported that he lives in the same house, upstairs.

Patient's Understanding of illness:

It was difficult to assess the amount of insight into her illness. She had difficulty concentrating and repeatedly reported that she had merely pushed herself too hard. She was tired and needed a rest.

It appears she is not happy being at the Waterford, would prefer any other hospital, but there weren't any beds available.

Nurse's Progress Notes:
May 15, 1995

She refused meds at 1800 hrs. Said she didn't like taking any medication. Judi was advised she was made involuntary, (certified under Mental Health Act) today and if she kept refusing her meds the doctors would probably order them to be given I.M. She said she would think about it, but did not come back looking for them.

She has been pacing the corridor and shouting. "O.J. is innocent, stating that she is "God" and was becoming more agitated. Kept repeating "oh my" "oh my" over and over. She repeats each sentence about three and four times.

She slapped two female patients. She said she slapped them because they would not listen to her. Dr. Hutchings notified and given Ativan I.M. at 2115. Back in her room at 2200 and has been sleeping since.

Judi spent quiet night on unit. Up a couple of times. Tearful, looking in the mirror saying "oh my, oh my" Returned back to bed shortly after. No more violent outbursts. Remains on close observation.

Social Worker's Screening Note:
May 16, 1995

Chart reviewed to determine need for social work intervention and to screen for family violence. It appears that Ms. Day has been admitted for psychiatric stabilization. Social Work intervention not indicated at present time

Plan:

Monitor through ward rounds.

I did not see a social worker for the five weeks of my confinement. Their job description in the Waterford Psychiatric Hospital must be different than I would imagine, especially since my mother had died during this admission.

When a patient is admitted to a psychiatric ward, the professionals ask the family questions about your behavior or personality traits. These answers are a part of the actual diagnostic procedure, but in my case, the diagnosis was already made.

The first question they asked my family members was: "Did I shop a lot?"

I was shocked when I read on my medical record's collateral history. "She, (meaning me) spends large amounts of money on clothes for her mother that she does not need." I really don't know for sure who disclosed this information to the doctors, I now believe it was my second ex-husband, but it was sure erroneous and it again portrayed the image that I inappropriately shopped, which is a symptom of "mania" to suit the diagnosis of manic depression.

If they had asked me that question, I would have replied. Sure I did. I shopped for my elderly parents. I shopped for my second husband who hated shopping. I shopped for my bed and breakfast that I ran. I shopped for my home and family, but I always shopped frugally and hardly ever inappropriately as I could never afford it.

My son questioned me once, thinking I was pregnant, when he found baby clothing on the upper shelf of my closet. I assured him that there were no babies coming, but when the little outfits were on sale, I would buy them as I never knew when I would need one for a baby shower at work, as there were over forty young women there working, and many of them were of child bearing age.

I had been going to thrift stores and buying my parents' clothes, especially Mom's, ever since I lived in Montreal in the seventies. Mom loved to dress well, and always wore beautiful dresses. I would go the Jewish Bazaars and buy her things at a fraction of the cost.

My mother would also mail me money in advance, so when I saw something that I thought she would like, I would buy it and send it to her. To help with our clothing budget, I would also buy items there for my children and me.

Since my mother had been losing weight and going down in sizes over the years, I kept her supplied with clothing that I bought in consignment stores and thrift shops. She loved everything that I bought for her and wore them with gratitude and pride.

Even my mother-in-law noticed how frugal I was with my hard-earned money. My shopping should never have been held against me to establish a diagnosis of manic depression.

Another statement on my chart taken from collateral history was that "I made mountains out of molehills." I knew exactly where that statement came from as my second ex-husband kept telling me that. In retrospect, the mountains were there, but I was trying to convince myself for years that they were only molehills

If I had followed my intuition more carefully, perhaps I would have saved a lot of unnecessary grief. For many years, I was suspicious of my second husband's cheating. My suspicions were confirmed when I physically caught him.

My sister-in-law told the doctor, and it was written on my medical record, that I had inappropriately bought a house, a Patty Duke move, which was inaccurate and far from the truth.

Karen, my daughter and I had purchased a home in Fredericton the autumn of 1994, when I attended university and we both lived in it together. After I became ill in the spring of 1995 and could not return to Fredericton the following semester, I made her the sole owner of the house, and she paid me back, in monthly installments the now badly needed money that I had invested in it. This money helped me with my financial survival when I was unable to work from most of 1995 to 1998.

Karen lived in the home until after she was married and had their son. They wanted to build a new home on the property next door, so I bought my dream retirement home back from them in the fall of 2003. They rented it from me, and kept living there while their home was being built, finished in the spring of 2004.

I moved back to Fredericton then and I am still living in my dream home today. What an investment I had made! It has helped both my daughter and me to be able to live in one of the most beautiful spots in Fredericton, overlooking the St. John River, with the walking trails just across the street, where we can watch the sunrise daily over the river, and the sunset in the stately trees behind our homes, where I have made the most beautiful gardens and walking trails that you can ever imagine.

I have the gardens lit up at night, so when I look out through my back patio door and kitchen window, I can always see that natural beauty until I am ready for bed. I am never more serene than while I am looking over my garden and watching nature as it unfolds each day. I have taken many photos of sunrises, but recently saw and photographed my first rainbow surrounding my home. It was phenomenal.

I want to share a few verses of a poem my mother wrote that will explain how I feel when I am in my garden.

My Cathedral

I don't need a great cathedral,
with windows of stained glass.
I find God here in my garden,
among the clover and the grass.

I don't need a great cathedral
With a pulpit built of oak.
As the walls of my cathedral
Are the spruce trees of my droke.

I don't need a great cathedral,
With a preacher loud and clear.
For nature has good sermons,
Which I am tuned to hear.

I don't need a great cathedral,
With a choir a thousand strong.
My song birds are my choristers.
They thrill me all day long.

I don't need a great cathedral,
Where strangers fill the pews.
So few of them are friends of mine,
That I have naught to lose.

My cathedral has a carpet
That nature did supply
And overhead a canopy
God's mural in the sky.

As I sit in my cathedral,
I worship God my way.
For peace of mind and happiness,
I thank him every day.

> I thank him that He's spared me,
> Though good health has passed me by.
> And someday I hope to meet him,
> In that home beyond the sky.

The only people, whom the professionals at the hospital interviewed to obtain collateral history, were my brother Ford, his wife Joy, and my second ex-husband. They did not even interview my son properly. This negligence is inexcusable and I paid dearly for their mistakes.

Clinical Clerks Notes
May 16, 1995

Talked to patient's sister-in-law who has known patient for a few years and describes her as very much goal directed, persistent, driven, and energetic.

She has done impulsive things in the past like buying a home in Fredericton. She has demonstrated behaviors in the past suggestive of persecutory and other delusions. She described patient as often getting very excited, talking fast, etc. She has seen patient twice in the past week and noticed increased symptoms, more energetic etc.

As far as relative knows patient has no underlying medical problems and has not been on any medication for her illness. They are unaware of any previous diagnosis, also unaware of any mental illness.

Family describes father and brother of patient as having similar personality – very energetic but feels it works to their advantage. Aware of Stressors

1. Mother's illness
2. Moving from Fredericton to St. John's
3. Mother's Day
4. Question financial problems?

Nurse's Progress Notes
May 16, 1995 1430

Judi remains agitated and tearful this morning. Stated that she was "crying for all the nurses in the province." Compliant with medications, with great persuasion. Judi reported that she has "racing thoughts." No other problems noted.

Clinical Clerk's Note
May 16, 1995

Saw patient this afternoon. Complained about being on medications, feels she does not need them and believes her thoughts are "jumbled" because of them. Also complained that she does not know what she says out loud and what she is thinking. Says her appetite and sleep are fine, but her energy is low since being in hospital.

She seems to attribute a lot of importance to her green flower. She finally told of an incident which happened last Sunday at church. A girl she did not know gave her this flower and passed her a heart shaped box. Patient believes this is a sign from God that she is to help all the nurses in the world to unite. She said in the past week she has interrupted sleep and this would go on all night.

Since hospitalization she has been quite loud, screaming Oh my God, Oh my God etc. when asked, she said she was saying this because she was receiving messages from God on how to help the nurses. She denies hearing voices from the outside. It is inside her head and she is convinced they are from God. At times she was heard saying she was God.

Mental Status Exam

Patient in bed when seen, fully dressed, fairly well groomed.
Complained of being tired "sleepy." Behavior appropriate, speech volume and rate normal.
Some flight of ideas and tangentially, but much less than before.
Labile affect. Denies hallucinations, but obvious delusion.
Grandiosity, religious delusions.

Impression:

Bipolar Illness. Mixed State

Plan:

No change in Management. Discuss at rounds tomorrow.

May 16, 1995 2000-2400

Judi spent a good evening on the unit. Very pleasant on approach, had a bath. Had her night time snack, and medications. No complaints voiced. She is presently settled to bed. Maintained on close observation.

He's the best physician who knows the worthlessness of most medicines. /Franklin

Waterford Hospital Admissions and Registration

Chapter 3

Normal Mental Status except no insight
May 17, 1995

Please note the bracketed sentences. According to Dr. Keith Pearce's medical report, they are indications, in my own words, symptoms of a major mood disorder.

The next entry in my journal was Wednesday, May 17, 1995.

I suspect now that my temperature is up as I feel so shivery. I asked the nursing assistant to take it for me. It was 37.2. That is high for me as my temperature is usually 35-36. I had two aspirins in my cosmetic case, which I took and am using Otrivin nasal spray to keep my sinuses from being congested.

Aspirin is the only anti-inflammatory drug that I can tolerate and I take it sparingly. I have had many bacterial infections, walking pneumonia, tooth abscess, sinus infections and my temperature has never gone up like most people's do. In fact it may go down lower.

I remember only once in my lifetime that my temperature shot up to 38.5. It was in 1985. I then phoned the hospital and staff health department to tell them how ill I was and that I would not be able to come to work or even go visit my family doctor as I was too weak.

I was told they would send a physician to my home, which they did. This medical home visit is recorded on my staff health records.

Neither bacterial nor viral infection was noted, but I lost my muscle strength and it took me weeks to recover enough to go back to work.

At that time, I had asked Dr. Jain if severe stress could do this to a person and he said "yes." This goes to show that "normal" should only be used as a setting on a dryer because each of us physiologically reacts differently to stressful situations in our lives, whether emotional or physical. This is why I believe that we must be treated holistically from head to toe when we are exhibiting any symptoms worth complaining about.

My mother and brother came to look after me until I recovered enough to help myself, and finally, after weeks, I got well enough to go back to work as usual.

Back to my journal May 17, 1995

(I want to write articles on nursing for the Telegram), like Florence Nightingale's book <u>Notes on Nursing</u>.

(I thought I would continue on with her <u>Notes on Nursing</u>: What it is, what it is not, and what it must become.)

(I will write ☺

(Nurse of thirty years admitted and committed to the Waterford Hospital as there were no beds available at the HSC or St. Clare's, after she had a psychotic reaction to medication, when she was quite sick with a heavy cold.)

(Judith Day, a clinical educator who is well known throughout Newfoundland for her contributions to Nursing, wants to let her fellow nurses know how frightening it is when one volunteers to be admitted to the Waterford Hospital) as she thought all the hospitals were under one board and the policies and procedures would be the same. Wrong!

I am locked up here and have no say into my medical treatment. I have to take what is prescribed by the doctors, or I will be injected. I am scared to death, and I hate the way the patients are ignored here.

(The nurses look very mean and are not counseling the patients at all. The patients are smart enough to avoid them because who needs mean looks? A lot of patients are coming to me for support and counseling.) Patients are confiding in me and seeking me out for support and understanding, even though I am so ill myself. It is such a shame that human beings are treated so badly by their supposed caregivers.

It is 0300, I am awake coughing and cannot go back to sleep. I have been sick now for three weeks with this illness. I cough till I turn my stomach and get a sharp pain in my head. I have never experienced such a cough in my life.

I am ashamed to say that these people who are carrying out nursing duties here are a part of the profession that I belong to and love. (Nurses must wake up if they are going to make a difference in Health Care Reform. Nurses' voices are beginning to be heard across Canada by the federal and provincial governments. Nurses have to be patients' advocates, and nurses must have a say in patients' treatments and education towards healthy living.)

Nurses should not necessarily follow doctors orders if they believe what is being ordered or done is negatively impacting the patient. If we are going to evolve professionally, Nurses must be more accountable and speak up when they see things that they know are wrong without the fear of losing their jobs. I know it is happening in peri-operative nursing and I am proud to be a part of that evolution.

(Peoples' health is wealth for the nation and everyone in it.

Illness is only wealth for the physicians and pharmaceutical companies.)

Nurse's Progress Notes
May 17, 1995 2400-0800

Judith slept well. She was awake on a couple of occasions. Maintained on close observation.

Clinical Clerk's Note
May 17, 1995

Patient discussed at rounds. There seems to be no question concerning diagnosis. Haldol will be increased and mood stabilizer added tomorrow. We will look for side effects.

Patient still has prominent psychotic symptoms, (I was having none) and so the neuroleptic will be increased since no side effects at present time. In view of non-compliance, Haldol will be given by mouth or intramuscularly.

Plan:

1. Increase Haloperidol 5 milligrams by mouth/im three times a day
2. Add Lithium 600 milligrams at night tomorrow.

Nurse's Progress Notes
May 17, 1995 0800-1600

She spent a quiet day on the unit, Very pleasant and approachable. Complained of racing thoughts periodically in which she hopes to change problems within the health care system. She denies delusions or hallucinations. Orientated times three. Memory good. No management Problems.

Clinical Clerk's Notes
May 17, 1995

Patient reports feeling very tired. Sleeping well at night. Remembers hitting a patient two days ago and says she apologized to her. Cannot remember why she did it and feels very guilty about it. Remembers saying she was receiving messages from God, but now believes they were her own thoughts. Content about being here, knows she is getting help.

Objectively she is well dressed and groomed, cooperative, behavior appropriate

Mood subjectively and objectively depressed. Tearful during interview at times, but smiling (appropriately) other times. Very pleasant. Speech rate and volume normal, slower than before. Affect appropriate. Orientated times three. Denies delusions and hallucinations. No evidence of either.

Impression:

Much improved.

Plan:

No change in management.

Nurse's Progress Notes:
1600-2000

Judi ate supper and spent remainder of her time in her bedroom.

2000-2400

Judi is much more settled although remains somewhat over talkative. She had visitor most of evening, so spent majority of the evening in her bedroom. She settled to bed at 2330.

Thursday, May 18, 1995

I slept off and on until six in the morning, and then I got up and took a shower. My head is so heavy with congestion that I do not feel very well at all, I am coughing continuously, which is really tiring me out.

I do not agree with the way that patients are treated here. (I have no choice in the matter to decide what medication I will take or not take as the doctors have me "certified," which means "detained against my will." I was told by the staff that if I do not take what is prescribed then it will be forced upon me by injection.)

(The drugs gave me a psychotic reaction and I made up my mind I am not taking any more. Most drugs have an adverse effect on my body.)

I was awakened by one of the patients screaming and I am frightened. I went to the non-smoking lounge area, but people are smoking in there as well. I have no place to go except my room as there is smoking everywhere. With my sinuses burning the way they are, it is very discomforting and painfully miserable for me. My coughing is getting worse and nobody is paying attention to it.

Clinical Clerk's Note
May 18, 1995

Judi reports feeling her old self today, only slightly tired which she attributes to her medication. She is compliant with her medications and has not caused any disturbances since the night of admission. She is no danger to self or others at this time. She is asking for decongestant.

Patient well dressed and groomed. She is cooperative and very pleasant this morning. She behaves appropriately, her mood is objectively and subjectively euthymic. No evidence of formal thought disorder. Her speech rate and volume are normal. No evidence of hallucinations, denies them. Also denies delusions. Says she knows the thoughts she was having were her own, not from God.

However, she is still wearing the green flower that she attributed much significance to in the past. She still feels she is here to rest and denies any mental illness, even in the past. Said ten years ago her admission was due to Gravol overdose which caused her to be hyperactive. I was also prescribed Halcion by Dr. Jain that added to the reaction.

She also feels like she should be here because she can help the other patients. It appears she is still having some delusions concerning her work as a nurse and having to help everybody. Grandiose.

Impression:

Bipolar-Manic

Plan:

1. Give unrestricted privileges
2. No decongestant at the present time
3. Continue present management

Relative (sister-in law) elaborated yesterday on an incident in 1993 involving patient's second husband. One and one half years ago it was uncovered that the man was having an affair with a younger man.

Apparently the younger man's mother called and told the husband one day that his boyfriend was with a woman she did not like. The patient's husband went and assaulted the other man, and so the charges were brought up against him.

The patient then went on to write repeated letters to Ottawa to free her husband to the point of getting letters back to stop writing.

It is unfortunate that when collateral history is given that it is not checked for accuracy before it is recorded on your permanent medical record and estranged spouses should never ever be interviewed as they may give inaccurate information just to be vindictive.

I have always believed in human rights and social justice. When I see violations and injustices, I act accordingly. I wrote letters of complaint to the RCMP Public Complaints Commission, after my second husband was injured by them during an arrest for "break and entry" and two counts of "assault," which were later dropped. He was placed in jail, and the police did not seek medical attention for him.

The police also told the lawyer, who I had arranged for him, that he was sleeping when the lawyer tried to contact him. Due to extreme pain, he could not even lie down, let alone sleep, as he had broken ribs that were diagnosed on x-ray the following day after he was released from jail.

The commission did an investigation, which found the police officers negligent and they were reprimanded. Both my second husband and I received from the Commissioner a letter of apology. My second husband could have sued for damages, but he wanted to keep the incident quiet.

Clinical Clerk's Notes, cont'd
May 18, 1995

As well she seems to have suffered a depressive episode at that time and was started on Prozac 10 mgms, which she stopped taking six months ago when she started school because she did not want to be on any medications that would affect her memory.

Yes, it was a normal reaction to the severe stress I was going through at the time and Dr. Jain recommended I go on Prozac 10 mg. I did not take the drug very long as I felt more sick and tired than depressed, and did not like depending on drugs. Besides I felt it was affecting my memory.

If Dr. Jain really thought that I suffered from bipolar disorder, according to[1] Compendium of Pharmaceutical and Specialities, CPS, he should never have prescribed Prozac 10mgs. without adding a mood stabilizer, as Prozac alone can precipitate a manic attack in people with bipolar disorder, which in my case in 1994, it did not.

I went through many other stressful situations in my life and did not resort to medication. I got through it on my own. I had a failed first marriage in 1973; I had a major automobile accident, involving numerous vehicles, in a tunnel in Montreal in 1976, and suffered head and neck injuries when I broke the windshield with my head that required over thirty stitches.

I was back to work within six weeks, as soon as I was off the crutches that I needed for contusions of both my knees.

At that time, I was dating a physician who admired my courage. I was thinking about buying a new car before I was discharged from the hospital. I was reassuring my co-workers and friends that I would be okay, instead of them reassuring me.

I was so thankful that I was alive and could move all my limbs, even though I could not bend my knees. I told them I would be back to work as soon as possible, and I was.

When I was out looking at cars with my friend, I wanted to go to the pound to see my demolished vehicle, which we did. The gentleman there made the remark when he pointed to my little white Datsun 1000cc Deluxe, "I don't know if the driver survived this one or not." I said "Oh yes she did!" "Here I am." He was amazed that I was out and about. I am a survivor!

I did not want my family, especially my mother, back home in Newfoundland to worry, so I did not even notify them that I was in a major car accident until I came home that following summer and my mother noticed the scars on my left forehead that were not healed as there were still residue of glass particles from the accident. I kept removing pieces of glass from my forehead for over a year.

I would never put any unnecessary grief and worry on my mother as she was dealing with her own chronic illnesses with symptoms of chronic fatigue syndrome that was never diagnosed.

Back to the Waterford Psychiatric Hospital:

Clinical Clerk's Notes, cont'd:
May 18, 1995

I will try to obtain information from her regarding follow-up with psychiatrist in the past and then get in contact with him concerning charts.

Dr. Jain's medical charts were illegible and impossible to decipher. Dr. Jain rewrote the records upon a court order during the malpractice trial. There was no definite diagnosis of bipolar disorder ever written in his records.

Affective disorder was documented as the discharge diagnosis in 1983 with the diagnosis of substance induced mood disorder, which can be one of three causes of affective disorder. I was definitely psychotic the first day of admission, but "recovered rapidly," according to Dr. Jain's testimony, which does not happen to patients who are manic or psychotic due to bipolar disorder.

Dr. Jain also stated in court to help establish that he had diagnosed me with Bipolar disorder in 1983, which he did not. "affective disorder" was written instead of "bipolar disorder" because "I put down, as you see the discharge diagnosis, I put down affective disorder." "It's a spectrum within depression." "The reason I put it Your Honor is this; affective disorder, it is a very nice little term which we used in Britain, and still use sometimes." "Affective disorder, which is a mood disorder,

---

[1] Compendium of Pharmaceutical and Specialties, CPS 2008, pg.1834

which could be mixed state, could be manic state, could be anxiety or mixed." "So that way, not every person to understand what—who this—what affective disorder is, so that was one intention, deter it, whenever I used to do it."—

The judge interrupted saying "to help out her job situation. Is that right?" Dr. Jain said "That's right, exactly." Judge finished his sentence "so that it wouldn't be affected." "Exactly, exactly" replied Dr. Jain.

I had a drug induced psychosis in 1983, not the symptoms of bipolar disorder and did not exhibit any symptoms of bipolar disorder for the next twelve years. Yet I was diagnosed as bipolar- manic upon admission to Waterford Hospital May 15, 1995 and forced on dangerous drugs after I was exhibiting normal mental status, except I had "no insight." I had no idea I was ever diagnosed with bipolar disorder and, in my opinion, I had led a productive, worthwhile life from 1983 to 1995, before 1983, and had no symptoms of that major mood disorder. Dr. Jain knew that despite his damaging testimony.

Dr. Jain's explanation of covering up the fact that I was diagnosed with bipolar disorder does not make sense because if he thought I did suffer from that serious mental illness, he should have told me and we could have discussed options for treatment. I knew and he knew that I had no mental illness that needed to be treated.

In the courtroom Dr. Jain went on to explain "psychosis" and stated "it is like pyrexia (fever) of the brain." "Psychosis could be organic, which you try to include the drug induced or metabolic induced." "Psychosis could be due to schizophrenia, could be due to depressive psychosis, could be mixed manic psychosis of mixed state or could be manic psychosis." "That's the way we go around this diagnosis."

Manic psychosis is the most severe form of bipolar disorder. I have been diagnosed with the most severe form of bipolar disorder and had a rapid recovery of the psychosis within twenty four hours by taking only one injection of Ativan. This does not make any sense to me as the recovery would have been weeks, taking antipsychotic drugs, if I actually had manic psychosis due to bipolar disorder and I would have continued to display symptoms if not treated.

Nursing Progress Notes
May 18, 1995 1400

Judi spent a quiet day; reports that her head is "clearing" and feeling better since she's been resting.

I have observed Judi taking control of the other patients, ordering them around and quickly going to their aid when they are upset. No concerns voiced. Stated that she wanted her Trinalin for her sinuses, but apparently this medication makes her quite hyperactive. Thus it is not ordered for her. No problems noted with her.

Judi spent a quiet evening. Up and around. Taking her meals and medications well. Complaining of blocked sinuses. She is very helpful with one of the older female patients (M.B). She had visitors this evening. Settled to bed at 2300 hours. No further complaints.

I swear to God, when I was given the psychotropic medications I would put them in my mouth, wait until the nurse went and spit them out. I had taken no medication since I was injected with Ativan the first night of admission and now four days have passed since I had been admitted.

Much are we beholden to physicians who only prescribe the bark of the quinquina when they might oblige their patients to swallow the whole tree. / Dalrymple

The entrance to Bowring Park, which is across the street from the Waterford Hospital, where many inpatients go on their outings.

Chapter 4

Bipolar Disorder vs. Reactive Psychosis

Friday, May 19, 1995

I have been listening to the radio and napping. I heard on the news that the government is setting up a national registry for blood donors. My argument is Why not set up a national registry for all health care recipients, with personal health histories recorded? I would not be in this situation if my health history were on a computer for Dr. Karagianis to access. He would not have had to receive information from my ex- second husband, which was wrong. I am trying to occupy my mind with good thoughts and have a little nap.

I had a nice nap after lunch. (My brother Ford visited me and we set the wheels in motion for me to obtain financial relief from Employment Insurance because I do not have any more sick leave accumulated.

I had used up all my sick leave during the fall and winter of 1993, when I was off work for over two months while feeling so sick and tired after my second marriage failed. I went through a battery of tests, even for AIDS, as the area near where my second husband was living was at that time considered the AIDS capital of North America. Thank God I was not infected! Living with an abnormally low immune system, due to undiagnosed fm/cfs/mcs, I was very blessed that I was spared from that fate.

Because of my spirituality, I learned to forgive both my husbands for cheating on me but I will never forget. I did not deserve to be treated that way and especially, in my second marriage being blamed for causing problems due to my suspicions of his infidelity, and my supposed lack of trust for him. He also told my family members about "my problem of mistrust" "paranoia" since he was such a loving husband and was being so faithful, until I finally caught him cheating on me.

After that I learned from my family members that he had phoned them, including Karen to apologize for his actions, which I found very noble and respectful. But no form of an apology would neutralize the deceit, thoughtlessness and disrespect that he had shown toward me as his wife. I left that marriage with the feeling that it is none of my business what other people think about him, me or the situation, and I moved on with absolutely no regrets and knew there would never be any reconciliation.

Sometime during that summer of 1994, (Dr. Jain had asked me if I would speak to support groups because my inspiration and courageous experiences could help others.), but around that time Dr. Jain moved away to London, Ontario, and at the end of August, I moved to Fredericton, New Brunswick to attend university.

I guess that was one reason why I headed the support group for people with fibromyalgia in St. John's after I was officially diagnosed with fm/cfs/mcs in 1997.

I expanded the group and renamed it the Fibromyalgia Self-Help Group of Newfoundland and Labrador and facilitated it for five years, until I moved to New Brunswick in the spring of 2004. I then became involved nationally as Director of the Atlantic Provinces for the National ME/FM Network Action Network, and I am still involved with that organization.

I had learned early in my life, while volunteering and helping others, I forgot my own troubles and then I could really help myself, in a way like nothing or nobody else.

One example of my helping others and my first positive experience with psychiatry happened twenty years before, in 1973 when I was a young mother with Karen and Brian, who were four and two years old.

I was invited for coffee one morning to a neighbor's house. She had a four year old boy and one year old twin boys. Both our husbands were away on business and she began to tell me that she hit her two boys in the face because they are not girls and have put them in the laundry basket and piled clothes on top of them to smother their crying as she could not tolerate it. She was shaking uncontrollably.

I knew I had to do something about this situation and convinced her to go to the emergency with me and she was admitted to psychiatry. I moved into her home with the five children, three of hers and two of mine, all under the age of four, and stayed there until her husband came home from his trip and made other arrangements.

She was treated and supported so well there, and was admired for having the courage to seek psychiatric help, and was praised for loving her children enough to do that.

She was hospitalized for a few weeks, came home with the support of her husband, family and a home care provider to help her with the children. This is an example of a positive psychiatric experience and every one benefited. The person is not labeled, not judged and not controlled, but treated respectfully and supportively, as I was treated by Dr. Jain ten years later in 1983, when I had a successful recovery from drug induced psychosis.

I don't know where these people are today as I lost contact when I moved from Montreal, but I heard years later that the family was doing well together. Raising children is difficult and it needs the help and support of mom and dad, the grandparents, friends, the neighborhood and the schools to ensure the best result is obtained. It has been said "it takes a village to raise a child." I believe that as well.

Back to my journal.

Friday, May 19, 1995

I have taken many knocks in my lifetime, but they have made me stronger. It allows me to appreciate every little thing I have and what I do for myself and others. (I like myself, I love my life and I will never allow anybody to ever hurt me that severely again. I have been cut to the deepest nerve and through the bone, but I have survived.)

Life is still beautiful for me, and I live each day to the fullest. I am happy and I am content, despite living with fm/cfs/mcs.

Friday, May 19, 1995.

The Clinical Clerk on Dr. Karagianis team spoke to me today about starting me on Lithium, and I asked "why?" (She stated because of my history of manic depression. I could not understand who formulated that diagnosis and on what grounds. I want to be drug free.) Dr. Jain who has treated me in the past knows how much I despise drugs. I want to be me, not something that drugs create.

Somehow I feel my estranged second husband contributed to this situation that I am in here and diagnosed with bipolar disorder. It is to his benefit that I am the mentally unstable one, as this would make him more credible. If this is the case, that is despicable and unforgivable. I have tried not to be bitter towards him for what he did to me, as that would only be self-destructive and I would have to use up the energy needed for self-healing.

My second ex-husband visited me while I was in the Waterford Hospital and he told me later that the doctors there had interviewed him and he wrote an affidavit to that effect. I have reasons to believe he also gave them wrong information that was also used against me in their diagnosis of bipolar disorder. He should never have been interviewed to obtain collateral history as we had already been legally separated for over a year and he had not seen me for over six months prior to his hospital visit.

Back to my journal

I know, as I have been told many times that I have a good, caring, sensitive, feeling personality. I am mentally stable, loved by many people, and I do not want to depend on drugs that I do not require. I can live without them beautifully as I have done most of my life.

During the past autumn and winter of 1994 and 1995, I attended university and took as many as seven nursing courses in one semester. I came through with flying colors, without taking any medications. I have won a scholarship from the University of New Brunswick's School of Nursing to return next fall and was also honored with a nursing award for excellence.

Taking seven courses was considered abnormal by psychiatry and that information was written on my chart. I was told by one of the nurses it was a sign of mania. I explained to her why I took seven courses, and also during the court proceedings, I put the explanation on record.

I had signed up for the seventh course as it was going to be dropped for lack of participation. My colleagues asked me to enroll and I did, with the understanding that they would do the leg work in the library, which was time consuming, to collect any information I needed. I would read and write the required papers for the course, which I enjoyed doing. My classmates and I both benefited: I received an extra credit and they had the opportunity to take the course that they needed for their degrees.

I received a beautiful card from my cousin Linda who has always been there for me, and many others in her family throughout my lifetime. She is one of the most caring and empathetic people I know. She spreads so much joy and strength to others.

The card contained the poem "Eagle when She Flies" by Dolly Parton

Linda wrote "May you always soar like the eagles." "Don't look at the past, just go forward." "Remember you are loved and cherished for the beautiful person you are, so don't ever change." "I will be there for you and I pray that God will bless you with Health and Happiness." "Just trust him." "Via Con Dios."

With all my Love—Linda

Friday, May 19, 1995

Dr. Karagianis interviewed me today. I really do not like his attitude towards me. He has a terrible bedside manner. He looks at me with such a mean look that he scares me. He misconstrued many things that I had said about incidents that had happened to me in church on Mother's Day, and misunderstood the significance when a child gave me a green silk rose, telling me that I believed I had special powers when I wore that rose. That thinking by him was totally wrong. I was spiritual, not delusional.

After talking to Dr. Karagianis, I won back my privileges and now I am allowed to go outside of the ward unattended. But I must stay here all weekend under supervision. I have not taken any drugs now for four days as I have been spitting them back out after the nurse leaves my room.

All my medication is officially stopped now.

A supervisor and colleague of mine phoned me from the Health Science Centre to say that she is trying to get a bed there for me. She is so good. She told me she will try to get me checked in under Dr. O'Loughlin, who is a female psychiatrist. I hope she is more understanding than Dr. Karagianis.

I am having a good day, but I wish the weekend would hurry up and be over with. There are many people who smoke here and it is really playing havoc on my sinuses.

One of the patients just gave me a blue floral ceramic heart. Since I have been here, the patients have given me a carnation, a bible, rosary beads, and now this beautiful keepsake.

I have kept that blue ceramic heart among my treasures since then and I often think about the person who gave it to me and wonder how she is now. I hope to meet her again one day.

Around three in the afternoon one of the patients came into my room and asked me to do up the fly in her jeans. She is trying to squeeze a size eighteen body into a pair of size fourteen jeans, and we did not have too much success. She finally gave up, removed her jeans and put on a pair of walking shorts.

I attended a session on leisure and recreation and the group made plans to go to the flea market on Sunday. I made several trips to the canteen and clothing store to pass the time. I bought an outfit for one of the patients whose clothes were dirty and smelled of perspiration. She was very thankful and pleased to receive them. I am good at picking out clothes for others, as for years I had shopped for my husband and parents.

My mind reflected back to the time when I was a student nurse and affiliating at the Waterford Hospital in 1965. I always felt I had to advocate for patients when I noticed that care was wanting, so I complained to the nursing instructor in the class about the way patients were being treated.

I noticed that they were given "routine enemas" twice a week which made them defecate (diarrhea) everywhere. The patients would be walking around with loose stool coming from them, but still they would have to receive these routine enemas. It was shocking and horrible to witness.

Anyway, that policy and procedure was changed after I spoke up in the classroom, and I have not stopped speaking up for patients since, especially when I see things that are detrimental to patient care, and could be changed.

Perhaps I have made some enemies speaking up for patients, but I am sure that some good has come from it somehow, and that is what matters to me.

Clinical Clerk Note
May 19/95

Patient seen today. She reports feeling well. Has a good appetite. Sleeps poorly because of sinuses and energy is lower than usual. She is convinced her illness is attributable to her flu, not a mental illness.

Well dressed and groomed today, pleasant. She is slightly irritable at times, but never out of control. Behavior appropriate, mood objectively and subjectively euthymic. Affect appropriate. Spontaneous smile, spontaneous conversation. No formal thought disorder. Only slight circumstantial, no tangentiality. Denies hallucinations. No evidence of delusions. No suicidal ideation.

Impression:

Diagnosis 1. Bipolar Disorder. Manic
**   2. Reactive psychosis**

Plan:

Discontinue all medications and reorder Haldol and Cogentin whenever necessary. Observe over weekend.

I talked to patient's brother today. Again he gives picture of a very impulsive driven person. She has done inappropriate things like calling her brother at 3:00 hours to go check on her husband.

That was the night my second husband was jailed after being injured, close to where my brother was living. It was a crisis situation and what I had done that night to obtain help for my husband was neither inappropriate nor unreasonable. I had no idea of the circumstances surrounding the incident with the RCMP at that time, and was only trying to get help for my husband who had phoned me from jail telling me he was beaten up by the RCMP and locked up.

She has spent a lot of money buying her mother new clothes.

That was inaccurate! I bought her "new" clothes at thrift shops and my mother paid me to do that.

She has given all the money she had in her purse to a collection plate, etc.

I have always emptied my change out in the collection plate as loonies and toonies are heavy in a purse. People with fm/cfs/mcs do not wear heavy clothing or purses over their shoulders as it causes more fatigue and pain.

Brother seems to suggest she has always "been on the edge."

Four days prior to admission he and a sister felt that she was again close to the edge"- Very high strung, etc. Brother will leave phone number.

My brother who is five years younger than I am was eleven years old when I left home to work and then I went to nursing school. We have never lived close to each other, not even in the same province most of our lifetimes. People are not the same with their level of activity, interests and accomplishments. What does "been on the edge" mean? The diagnosis of bipolar disorder had been established upon admission, and then incidents and expressions recorded to fit that diagnosis.

My only sister, who was referred to above, lived in Hamilton, Ontario, and I had not seen her either for over a year. My sister and I had never lived in the same province since she left home at age fifteen to study in St. John's, and then she moved out of Newfoundland.

It is amazing how the diagnosis of bipolar disorder was made and then incidents were written on my chart to substantiate that serious diagnosis.

My family's descriptions of my personality, which were recorded on the progress notes as abnormal, drove wedges in our relationships that took years to mend.

Nurse's Progress Notes

May 19, 1995. 1330

Judi spent a quiet day. While speaking with her she believes that she has been given the wrong diagnosis. She says she is not manic depressive and will not take her Lithium. She says that her activity level was due to her antihistamine. She is complaining of congestion, sore throat and coughing. She is going on a possible afternoon pass with a friend to file unemployment insurance for long term disability.

2000 hours

Judi has been up and about all evening. She refused her medication at 1800 hours and says that she would rather take it going to bed. She is pleasant and cooperative.

2000-2400

Tearful on occasion, settles fairly easily. She is in her room writing, showed some of her writing, had some rambling at times.

Judi says she is afraid she'll be locked here forever if she "goes psychotic again" Says she got ill prior to admission due to long bout of flu, not eating or resting and taking Trinalin.

Says she doesn't need any medication like Haldol. She is complaining of feeling febrile (hot). Temperature was 37. Said that was extremely high for her. She was reassured by writer, with no result.

A rule of thumb in the matter of medical advice is to take everything any doctor says with a grain of aspirin /Ace

The famous Swans of Bowring Park

Chapter 5

The Infamous Weekend

Saturday, May 20, 1995

The nurses are not nice to the patients. I went to the station to see if Mrs. B could have her bath as she is in pain. I called out "hello" as there was nobody in sight. The response came "WHAT" in a horrible tone. I asked if the person could please come to the wicket and speak to me, which she did.

One might say it was none of my business to meddle into other patient's affairs, but you would have to be there yourself to witness the neglect to understand why I being a clinical nurse educator felt I had to intervene for them and be their advocate as the nursing care was so wanting and neglectful, it was too painful for me not to interfere.

Things had not changed much since I was here affiliating as a student nurse in 1965. The patients were ignored just as much in 1995 as they had been thirty years before. When I was affiliating there as a student, it broke my heart to go to work every morning and there were afternoons that I could not even go back. At that time, after I complained about the nursing care on the geriatric ward, there were supposed changes made. I know for sure about one change and that was, student nurses no longer had to go to that particular ward.

Back to my journal:

There are no smiles, no willingness to help, nothing from the nursing staff.

I asked if Mrs. B. could have her bath and the nurse replied that there was nobody to watch her. I volunteered, but a nurse or female staff member must accompany her. Mrs. B went back to bed until after breakfast. It appears that it is too much trouble for the nurses to do anything.

During breakfast I sat with three patients, one was complaining of a headache, one a backache and the other was angry at the world, so I kept saying to the three of them that they will feel better again someday.

Mrs. B. would like to have her pain pills before she gets into the bathtub, but nobody is moving to give them to her. She came into my room because she saw a cart in the corridor, and she thought her pain pills were coming. I informed her that the cart she saw was the housekeeping cart and I told her she should go to the desk again to get her pain pills. It has been over an hour now since she told me she needed her medication. I have never seen patients so ignored in my life.

The nurses are always in the glass enclosure called the nursing station. They rarely ever mingle with the patients, not even in the dining room during meals. The patients are handling hot kettles while getting themselves tea and coffee.

I have tried to help the older patients with their hot tea and coffee as it is very difficult pouring from that huge hot kettle with a big spout into Styrofoam cups, which tip over very easily as they are being filled. I thought "this system is absolutely cruel"

Mrs. B. has just come into my room again begging for her pain pills. I again went to the nursing station to 'please' give her something for pain. Her son had told me she has three degenerative discs in her lower back and osteoarthritis has set in. She has been suffering for hours and it appears that the nurses are oblivious to patient suffering.

I have coughed so much that my head throbs. I become nauseated, my chest is sore, yet not one nurse has commented on the cough or has suggested that they obtain an order for some cough mixture. I cannot believe that these people belong to the same profession as I do. They have their heads in the sand and are punching in shifts. They display no concern whatsoever for patient well being and safety.

Finally at 0945, Mrs. B received her pain pill. I tucked her in bed and hopefully she will nap (without her bath). Her comfort is more important now then her cleanliness.

At 1100, I had a discussion with one of the nurses about myself and my diagnosis. I am going to let her read my writings. I have given her a couple of papers that I have written on nursing. I will let her read what I have written today.

In retrospect I believe that was a big mistake.

I phoned my friend and she told me she was going to come see me after her daughter's gymnastics display at the Avalon Mall. I wanted to go with her to the display as I like gymnastics. She asked me to find out if I could go and that she would call back in an hour.

I asked one nurse, and she said "definitely not" as I was under twenty-four hour surveillance. I asked her "since when?" I had permission yesterday to go the employment office downtown with my brother. She told me she would check and let me know. She did not get back to me.

After an hour, my friend phoned again, I spoke to another nurse about the afternoon pass and she told me she would check for me after she finished giving out the medications. I informed her that my friend was on the phone and that I had asked over an hour ago.

The nurse said she would have to phone the doctor to get an order. I replied with a question. "Who knows more about my condition right now, the nurses on the floor, who have been observing me, or a physician who I have not seen for over twenty four hours?" I asked her why she could not make an intelligent decision and recommend to the physician that it is safe for me to go out to the mall for a couple of hours.

By this point I was definitely "mud" in the nurses' eyes. I guess I was looking for a professional response that she would phone the doctor, relay to him that I was mentally stable and reliable and would recommend that I be allowed to have another pass, as I did yesterday.

Nurses are so controlled by physicians and are led to believe they cannot think for themselves. Nurses are supposed to be educated professionals who are accountable for their actions and yet they need to get an order from a physician (who has not laid an eye on me) to allow me a pass that the same physician granted me the day prior, so I could go downtown to the employment office to sign documents. Would that pass be granted yesterday or would I be allowed to sign documents, if there were any doubt about my sanity? These are strange happenings.

Eventually I was allowed to go out to the gymnastic display with my friend, but I was very weak, tired and not feeling very well at all, but still enjoyed the freedom and the display. My friend testified during my medical malpractice lawsuit about that afternoon and how pale, fatigued and ill I looked because of my upper respiratory infection.

A few days before my admission, I had been prescribed antibiotics with the Trinalin, but after I was admitted to the Waterford hospital, the antibiotics were also discontinued without completing the full dose.

Nurse's Progress Notes:
May 20, 1995 0000-0800

Awake several times during the night. No management problems.

0800-1600

Judi is more settled today than previous days. She spoke at great length to writer this morning. She informed writer she is not manic depressive and feels this admission is due to taking Trinalin tablets and the many stressors present in her life.

She claims that Dr. Jain informed her she has done "marvelous" considering all the stressors in her life. Patient is very fixated on nursing issues and her responsibility in her work place. Mood is labile. (Lability denotes free and uncontrolled expressions of the emotions, emotionally unstable.)

She is tearful when discussing her marital situation. However, in conversation Judi has no evidence of pressure of speech, able to be interrupted during conversation.

She informed writer that she is going to write articles for the Telegram on nursing issues in order to "make the public aware of the changing health care system".

While on the unit, patient is helpful with other patients. However at times she is interfering with other patient's treatment; i.e. coming to the nursing station several times informing staff nurse that another patient needs her medication at a specific time in order for the patient to 'start her morning care.'

Patient was spoken to with regarding her interfering behavior and accepted this well.

Patient requested accompanied pass to visit her friend this afternoon. Dr. Hutchings (doctor on call) was notified and pass granted.

Patient is refusing medications of Trinalin as she states this caused her "to become sick". Patient also informed writer she was taking "Prozac" which was prescribed by Dr. Jain. She also gives a period in past of not being able to sleep, because these times she stated she took "Gravol", which made her "hyperactive".

The Trinalin that I have been saying caused the psychotic reaction in the first place was ordered by Dr. Karagianis, one by mouth, twice a day, on May 19, 1995, but I refused to take them, so they were discontinued on May 23, 1995. I do not understand why Dr. Karagianis ordered Trinalin after I kept saying that this was the drug that I believed I had the reaction to. Dr. Karagianis did not pay attention to anything I was saying or what the nurses were writing, and forged ahead with his misdiagnosis of bipolar disorder and maltreatment.

The first day of admission, when Brian came to see me, he had brought into the hospital the huge bottle of Prozac that I had at home. This information was not recorded on my medical record until now.

For a couple of weeks prior to admission, I had been taking Prozac 10 mgms. as I was so sick and tired from the upper respiratory infection that I had been battling for weeks. I thought taking Prozac would help me with the fatigue as I was beginning to feel depressed about not being able to continue working as I had no sick leave. I definitely needed to be gainfully employed for the summer if I were going to be able to go back to New Brunswick and finish my education in the fall.

I was always pushing myself to the limit to do the things that I wanted to do, and this was another example of how I dealt with fm/cfs/mcs. I was experiencing different symptoms than people do with bipolar disorder and perhaps, if the health care professionals paid a little more attention to what I had to say, I could have been properly diagnosed during this admission to hospital.

I had not taken any Gravol since 1983, when I was told it possibly was the drug that I reacted to in combination with antihistamines, (drugs to neutralize and antagonize the action of histamine for treatment of allergies and stops runny noses) and decongestants (used to reduce congestion or makes it easier to breathe).

Nurse's Progress Notes, Cont'd
May 20, 1995 2000-2400

Patient is spending time writing in her room this evening, keeping a low profile. No complaints this evening concerning increase temperature. No attempts to provide care for patients this evening. She settled to bed after 2330 hours.

Sunday, May 21, 1995 at 0430.

I must write about C., one of the patients here. He has a very sore leg as a result of a fall from a bridge. His leg is swollen, reddened and itching. The inflammatory healing process is causing him discomfort. He came into my room last night scratching at the necrotic scabs that have developed on his shin bone. They itch and burn which causes him to pick at them.

I noticed that one of the scars was draining pus, and I fear that infection is setting in. I know how serious infection of the bone is and it has to be treated properly. Last night I spoke to someone at the desk, thinking it was a physician or a nurse, but it was an orderly. He offered to remove him from my room. I said I am fine with his being in my room, but he needs medical attention, and the orderly left.

I encouraged C. to keep his leg moist and I applied some aloe vera cream, which is very soothing and wrapped his leg in cold wet facecloths. This helped him from scratching it. I encouraged him to keep his leg elevated and stay off it when possible.

I spoke to the nurse about C's leg needing medical attention and she became very angry at me for interfering in patients' treatments. I spoke my mind, telling her that C's leg needs medical treatment and asked her to get some flamazine cream or get something ordered to take the itching and burning out of his leg or else he will be scratching it, and it will become more infected, right to the bone.

I cannot believe the lack of nursing care that I see on this unit! Why don't the nurses listen to the patients and talk to them when it is needed? They ignore the patients completely. The patients fend for themselves. What a waste of money on nursing staff.

At 1030 I went to church with two other patients. The psychiatrist on duty wanted to see me at the same time, so they sent a staff member down to the chapel for me to return to the unit. I was disappointed that I missed some of my singing, but it was important for me to see the psychiatrist because I wanted to get a pass to go out with my father in the afternoon.

I know I spoke very intelligently to the doctor, but he had to leave before he was finished with my interview, as he had an emergency on another unit. I assumed there was no reason for him to refuse my afternoon pass, so I was just waiting around for Dad to come so I could leave with him.

About five minutes before Dad was expected, I was informed that I was not allowed to go out with him. I was very disappointed as there was no logical reason why. Dad arrived on the unit and we talked, but I was forbidden to go with him for lunch.

I was crushed and started shedding tears. After all, I was allowed out yesterday with my friend and I was perfectly fine mentally today. What could be different now?

I gave Dad the journal that I had written for the past week and asked him to take it to my former brother-in-law who was a lawyer and my children's uncle, a man who had always shown the utmost admiration and respect for me. I thought if anyone could help me he could.

I told Dad to inform him that I needed two independent psychiatrists to interview me with a lawyer because I know I am mentally competent and have never been diagnosed with manic depression, and definitely should not be forcefully medicated with harmful drugs. Something is seriously wrong and I have to get out of this trap.

When it was time for Dad to leave the unit, I walked down to the car with him and he gave me a poem that mom had written for me since I was admitted here. He told me that my mother was in hospital again and that she would probably need oxygen when she comes home, as she finds it hard to get her breath. Mom is fading, and the stress of my being here is not helping her. She loves me dearly and prays for my happiness so much that it upsets her when others do not understand me the way she does.

If she dies before I get a chance to see her again, I will always believe that the grief that this admission has caused her brought on a heart attack. I know her heart muscle is weakening and she cannot cope with the increases in blood pressure aggravated by the added stress. This crisis is killing her and there is absolutely nothing I can do about it as I am locked up in here.

I hope and pray that when she dies, I have the rationale and emotional intelligence not to blame myself for her death. My siblings may, as we know my situation is killing her, but that guilt would destroy me.

The poem Dad brought in to me was the last poem my mother wrote.

"FAITH"

If a radio's slim fingers
Can pluck melody from night
And waft it o'er a continent
Or ocean.

If the soft notes of a violin
Can be heard above the city's din,
And bring sweet melody
To our ears.

If Crimson break of dawn
Can set the sky on fire,
And the world be filled with singing
From God's Feathered Choir.

Then why should mortals wonder
That God hears prayer.
Keep on praying, Judi darling
All will be well.

Sunday, May 21, 1995

Dr. Jain had been following up with me since 1983, especially this past few years and has never told me that I had a mental illness, just anxiety when I became overwhelmed with chronic fatigue and stress. Many times it was written on my records "denies being depressed." I did not feel "gloom and doom," just fatigued and physically ill with one infection after another that contributed to the feeling of malaise that required me to rest more than most people.

Unfortunately Dr. Jain is not in Newfoundland anymore, but my charts must be somewhere. I have been described as a sensitive, loving, giving person, whose biggest problem is that I try to extend myself to help many people. Finally I have learned to say "no" when others seek out my help.

I was feeling very positive about my future before I came in here and wanted to write educational material. I love writing and one of my university professors had encouraged me to pursue a writing career and apply to the government to write for them when I returned to Newfoundland and Labrador from New Brunswick.

My enthusiasm for my professional career and my carefree lifestyle, without child raising responsibilities, as my children are adults now, make me content and happy. I should not have any worries financially now as I have both my children raised, educated with university degrees and situated in their own homes.

But right now I am in a mess, as I am unable to work and support myself, or do anything as I am locked up in a mental institution and forced to take drugs against my will. I am losing control of my autonomy and my life.

I walked out to the main door of the ward, which was open. Usually it was locked, and the thought came into my mind of running out, down the stairs and being free. However, very quickly I realized that if I did that, I would really have something on my file that would be held against me.

I got that idea out of my head immediately and walked up and down the corridor, trying to relieve some of the tension that had built up since noon, when I found out that I was not allowed to go out for the afternoon with my father.

I retired to the lounge after a social gathering in the eating room. Tonight was the first time that I went into the lounge as I could not tolerate the smoke before. With my head cold clearing more daily, I am beginning to tolerate the smoking. I cannot believe that.

I spent some time talking to some of the friends that I have made while confined in here. We listened to V.O.W.R. church service on the radio, and sang hymns together in my room. Then I took a relaxing bath and got ready for bed around 2000. I had had a long day and so I was ready for sleep around 2100.

My mother taught me how to fall to sleep naturally and quietly by saying. "Dear God" (take a deep breath) "please clear my brain and help me sleep." Repeat the above as many times as necessary. It really works for me a lot better than counting sheep, but tonight, even though I was exhausted, sleep was not coming.

As I lay in bed I started to cry. I had to stop that as I know how destructive prolonged crying is for my body. Crying at first is a good release, but after only a few minutes, the chemical make up of the tears changes, and it is no longer therapeutic, but damaging.

There were times in my life I knew I had cried for hours, especially after the break-up of my first marriage. I had little support or empathy from my family as divorce in the seventies was very rare, so people kept their marriages intact and lived unhappily ever after. People, who did not, were a disgrace to their families, despite the circumstances for the break-up.

My marriages were sacred to me and I took my vows very seriously and remained faithful. Being married was forever, despite its ups and downs. I believed I was happily married and I thought my husbands were as well.

My first husband would tell me "I couldn't imagine loving you more than I do, but then I do." That changed when we moved to Montreal. During that time, I had told my friend Rosalind that I trusted my husband completely and that I believed he would never, ever cheat on me. She stated, "Judi, I would never trust any man to that extent." I replied "if I couldn't trust him, how could I love him?"

I saw the movie <u>The Truth about Love</u> lately and it brought back memories of what my friend had said to me when I was so naive and vulnerable. I have learned though, that when the trust went, the love went as well, and as hard as I tried, it did not return. Our never ending arguments had turned me into an unhappy, frustrated, angry, and nagging wife as our marriage kept crumbling. Without love and trust, I could never live in a marriage and be content and happy. I have never regretted my decision not to reconcile our differences and stay married, as how could I trust him again? But I have regretted that my first marriage failed and we would not grow old together with our children and grandchildren.

After listening to many testimonials of extramarital affairs, this is exactly what actually happens when a married man gets involved with another woman, or vice versa. It destroys everything! Also, I do believe that what a person does with you, he is capable of doing to you as well.

Back to my journal:

Sunday evening, May 21, 1995

I thought of my mother in the hospital, and wondered if Dad had arrived home safely, so I phoned him at home.

I told him that I believe that it was my second ex-husband who had misinformed the doctors and told them that I had been diagnosed with manic-depression and that is why they are going to medicate me with Lithium.

(I will never want to lay eyes on him or Ford again. I believe now that they were instrumental for my diagnosis of manic depressive illness by giving inaccurate collateral history.)

Until I obtained my medical records from the Waterford Hospital in 1996, I had no idea that the mistaken diagnosis of 'bipolar-manic" had been written by the ward clerk on the transfer form and "letter" that had accompanied me to the Waterford Hospital, May 15, 1995.

After that, my family members were told my diagnosis was bipolar disorder and they were questioned about my behavior, thus giving examples that were taken out of context to fit the diagnosis of bipolar disorder. By then, I was taken completely out of the picture, and being treated like an uncommunicative animal, which was very troubling for me.

Nurse's Progress Notes
Sunday May 21, 1995 0000 -8000

Slept in naps, she was awake at 0300. Pushed into nursing station at 0400 hours, demanding cream for male co-patient. Said she had put Aloe Vera cream on his leg earlier. Writer requested she not repeat this as there might be perfumes or anything in the cream she had used.

Judi became more upset. She said that Aloe Vera was a 'natural healer'. She said she had been a student at this hospital in 1962 and had reported problems here and changes had been made. She said that she was now a nurse and an educator, and was keeping notes on everything that was wrong here and things would again change.

Settled after some time and returned to bed. Awake at 0630 hours and requested a shower.

0645

Requested shower at 0630 hours. Explained to Judi that showers start after 0830 hours, normally unless person soiled during night. Judi insisted on taking shower as she has had shower past three mornings at this time.

Judi asked writer if nurses on this ward are registered nurses as a registered nurse can change any rule or guidelines that she wants. Judi says a registered nurse is boss and has to answer to nobody. The nurses in this hospital are not registered nurses as they don't make their own rules, but just follow rules according to Judi.

She is very abrupt and confrontational upon approach.

This writer spoke with patient this morning regarding her interfering with patient treatment of others on the unit. Judi stated that she was a nurse and understands the needs of others. Patient informed she was presently a patient on 43B and not functioning as a "nurse".

Writer attempted to explain to patient the need to take medications as her mood was escalating. Patient insisting she did not need medications and that they would make her "psychotic".

She became very confrontational during the conversation, informed writer she is recording all interactions going on since her admissions and will report them to others in authority.

She states she is being treated "unjustly", however unable to support same.

She also stated she does not "trust the nurses here". Informed writer she will call her lawyer and nursing supervisor from the Health Science Centre in order to "get out of here."

Patient completed Review Board Hearing Papers.

Patient requesting pass to go with father this afternoon.

The nurse informed me about my right to appeal my confinement so I completed an Appeal for Release application form and gave it to her. I learned after I obtained my medical records in 1996, that the document was sent to hospital administration, but was not acted upon. This I believe was in violation of the <u>Mental Health Act.</u> I wrote this information in my statement of claim for my medical malpractice lawsuit.

During that lawsuit Dr. Karagianis testified in the courtroom that he felt that my hand writing on the Appeal for Release form was abnormal as it displayed "hypergraphia," or I had written too much information on the form. I was only trying to explain everything in detail so that the officials would understand that I was not mentally ill and did not need to be institutionalized against my will.

Dr. Karagianis testified that after reading the Appeal for Release Form, it convinced him more than ever that I was manic and in need for treatment.

The judge grinned after he took a look at the form and stated "you should see some of the forms I get with traffic violations." What a joke! They all laughed except me. There was nothing for me to laugh about.

Nurse's Progress Notes,
Sunday, May 21 0800-1600

(Doctor on Call) notified and visited to assess. See Note
Physician: Seen today.
Still uncooperative, circumstantial, and not sleeping well.
Hypomanic (question)
Impression: bipolar disorder
Management; Restart Haldol
Patient to remain on unrestricted privileges and advised to discuss passes with her doctor.
Patient's father (Mr. Sparks) visited the unit this afternoon. Informed writer he knew Judi was not well for some time. Wants to "take on the world" and states Judi is always writing things in order to "fix" them.
Judi tearful during the visit and insisting she did not need to be in hospital because she is "sane." She asked her father to call her lawyer in order to get her out of hospital. Patient secretly placed written notes into his jacket pocket in order for him to deliver to her lawyer.
Patient was angry toward staff when pass was denied. Mr. Sparks informed writer he was worried about how to handle her while out on pass. Father was in agreement on pass not given.

Stan Power had driven my father all the way from Sibley's Cove that morning to spend some time with me, and three of us were very disappointed that it could not happen. Stan wrote this letter during my medical malpractice lawsuit.

I Stan Power of Caplin Cove, a friend of Clarence Sparkes and Mrs. Louise Sparkes. I knew Mr. Sparkes for forty years.

One Sunday in May, 1995, I went to St. John's with Mr. Sparkes to visit his daughter Judi Day who was in the Waterford Hospital as a patient.

I stayed in the car in front of the main steps and Mr. Sparkes went into the hospital to get Judi.

A while later Clarence and Judi came out and down the steps. He told me that Judi could not come out with us for lunch as she is not allowed to leave the hospital.

I spoke with Judi and she appeared as mild mannered as I always knew her, and we could see no reason why she could not go.

Clarence did not tell me that he did not or could not take the responsibility for her.

Judi and Clarence sat in the car for awhile talking and I walked a short distance to have a smoke and let them have some privacy.

Stan Power
Witnessed: Mabel Nixon

Normally I would read myself to sleep and I love researching topics to write about. When I wake up during the night with ideas for papers, I would get on my computer and enter my thoughts, then I would go back to sleep.

I cannot imagine "that" being abnormal because I received such high marks on my papers, 29/30 on some. I had very few papers on which I received a mark below A. Is that wrong?

My mother told me she did the same thing with her poetry. She would wake up with an idea for a poem, write it down and then go back to sleep.

I have never had any problem sleeping 8-10 hours a night. I hardly ever sleep during the day, unless I am physically ill with upper respiratory infections.

I am constantly being awakened here because of people screaming and then I read and write until my fingers become tired and I get sleepy again, so then I can settle down to sleep.

(I do not want to change my zest for life with pills. Why should I? Before this admission, I exercised, walked, sang, did gardening, wrote, read, watched TV, and read the newspapers daily.

I happened to catch the virus that was going around St. John's when I came home from Fredericton and had a reaction to the medication I was prescribed. Is that a reason to drug me with dangerous antipsychotic drugs?)

To keep myself occupied to reduce the boredom here, I walk the corridors, sing, write, and watch T.V. I cannot read anything as my concentration will not allow me that privilege. I have studying to do for my certification exams, but I cannot even read the newspaper.

If I were not so courageous and strong, I would never be able to tolerate this unjustifiable confinement for so long. My faith in God has helped me through one of the most difficult weeks that I can remember in my life.

The stressors here in Newfoundland and Labrador are too overwhelming for me and I know now that I am strong enough to move away from people and an environment that are not conducive to my well-being. Life is too short to waste another minute on this matter.

God has given me the serenity to know I cannot change this mess. I must accept the defeat and move on for my own survival and personal growth.

I had a conversation with one of the nurses tonight and I believe that I have convinced her of what my estranged second husband has done. I am confident that she believes me and I am sure after I speak with Dr. Karagianis that he will release me from here.

I have learned one lesson while locked up in here and that is there is strength in prayer and trust in nobody except yourself and God. This confinement has taught me that innocent people suffer daily because of other's careless actions.

Progress Notes, Sunday:
May 21, 1995 1600-2400

Patient more settled this afternoon. Refused PRN (whenever necessary) medication of Haldol when offered. She is adamant about not needing medications.

She continues to write notes in her room. Calls same her book and hopes to have it published when discharged. She is showing some of her writing to co-patients at times. She is spending long periods of time on the telephone during the evening

She is less confrontational and abrupt this evening than has been recently. Settled to bed at 2300 hours.

Awake at 2400 hours, came to nursing station to inform staff she plans to have a lawyer and stenographer come to the unit Tuesday for a meeting with Dr. Karagianis to change status of Involuntary.

Judi was informed that her Review Board papers are the correct route to take concerning status but she feels that will take too long.

She wants an interview with Dr. Karagianis and her lawyer and same taken in minutes by stenographer.

Judi then said her second husband who she is now separated from has told lies about her to Dr. Karagianis and because of this was certified. She spoke with writer then for two hours about life over past one and a half years. She admits to being depressed, but it was circumstantial over marriage breakup.

She speaks of all her plans in the future to inform public and nurses of the changes about to come with health care reform. She plans to continue education to Public Health Department level and doesn't care if she becomes penniless doing same. During conversation speech, tone, rate normal. Goal directed conversation, spontaneous and affect appropriate.

She spoke of "green rose" and how a child on Mother's Day gave it to her and how she was touched by this.

Judi maintains she has had another psychotic episode to Trinalin and that 'manic depression' is not her problem. She was given assurance she will have ample opportunity to speak with doctors this week. Settled back to bed at 0200

One of the chief objects of medicine is to save us from the natural consequences of our vices and follies./ Mencken

Bowring Park near the Greenhouse

Chapter 6

Frustrations Mounting

Monday, May 22, 1995..

    I slept until six in the morning and the first thing on my mind when I awoke was my mother, who is hospitalized in Old Perlican. The worry of my being sick has caused her blood pressure to strain her damaged heart, and according to the nurse I spoke to at the hospital, she is in congestive heart failure. Lasix is decreasing her plural edema, and Ativan is keeping her sedated. I know that my mother will not be able to tolerate too much Ativan, a relaxant, as she is as sensitive to medications as I am. I hope and pray that nothing happens to her until I get released from here.

    I know she is worried about me, so I asked the nurse looking after her to tell her that I will be released tomorrow, and I will drive over there to see her then. That will keep her going until Tuesday. If I don't get released by then, I will say I made a mistake and I'll be there Wednesday. She must have something to hang on to now. The sedation is going to decrease her respirations and God knows she does not need that.

    Mom deserves to be in her garden today, not worried about me and sedated, especially since I am okay, except that I am locked up here without just cause.

    I took my bath before breakfast and dressed for the day. I planned to spend some time outside, so I dressed for walking. I walked for thirty minutes before breakfast and arrived back as the food trays were being given out.

    I could not phone Dad until the phone room was opened, which is usually 0900. I knew he would be up at 0800 as usual, but I had to follow rules and regulations and wait until 0900.

    I stood by the door of the phone room at eight fifty, but the door did not get unlocked until nine fifteen. Those twenty five minutes felt like two hours. The nurses just don't care and if I complained, it may not get opened for another twenty five minutes. By then, Dad was out in the garden, as usual, doing his work. He was glad to hear from me and we chatted for a half an hour. I assured him I was fine and hopefully, I would be home tomorrow.

    The psychiatrist on call would not even speak to me as I had requested, so I cried alone for awhile, thinking of my helplessness and how uncaring and inconsiderate everyone has been to me since I have been here.

    To pass the time until lunch would be ready, I made a few more phone calls. My appetite is decreasing, but that is okay as I need to lose at least ten pounds.

    After lunch I went for another walk to get rid of the stress caused by this confinement. I feel like I am an innocent person being imprisoned, and I have to behave myself so I can get my parole. I have been misjudged and sentenced without hearing evidence against this serious diagnosis. Does everybody have to fit into a conforming mode of behavior that is considered normal and ideal by taking drugs? When I am properly diagnosed by concrete evidence of manic depression, then I will, and only then, take whatever medications are required.

    My father is energetic, loves gardening, makes interesting plans, builds a museum, writes his family history, and is quite active for his eighty two years of age. I wonder if his behavior and zest for life was considered abnormal or if he had been drugged, would he have had such an interesting life? I doubt it.

My parents are two of the hardiest and most interesting people I know. I have great admiration and respect for their accomplishments and their never ending love for each other, despite their hardships, grief and, losses that come naturally while living together for over fifty years.

This is a Poem that my mother wrote to my father.

Our Lives Together

A long time ago fate caused us to meet
And a friendship began so loyal and sweet.
But as the years passed, into love it did grow
Growing stronger and stronger, as days come and go.

After three years of courtship so pleasant and sweet
We decided to marry to make life complete.
My answer was "yes" and so happy were we
Altho' we knew not what the future would be.

The smallest of diamonds I so proudly wore
If it had cost millions t'would have pleased me no more.
And our vows at the alter so sacred and old
Were sealed by the tiniest ring of real gold.

God blessed our union, I know he did care
For he sent us an angel, our happiness to share.
For reasons unknown, she'd not come to stay
On a cold winter's morning, God took her away.

Our hearts were so broken, we couldn't understand
But there's always a reason for God's perfect plan.
He gave five more children, two daughters so dear
Three sons strong and handsome, who really do care.

No lives can be perfect, things often go wrong
But we overcame them, our love is so strong.
That great word forgiveness, the greatest in life
Can fix all the problems, twixt husband and wife.

My real trouble began in year '83
I was struck down with cancer as low as could be.
Dear, how you stood by me, by day and by night.
Your strength ebbing through, it helped me to fight.

God knows how I've suffered, these recent two years.
With strokes so depressing that often bring tears.
I've felt helpless and useless, a burden to you.
But your help so sufficient in bringing me through.

But God up in Heaven has blessed us with wealth
Not counted by dollars, but love, contentment and health.

A home that we cherish, and children the best
Our beautiful garden, and trees where birds nest.

We've blended our lives for fifty three years
Some happy so happy, some damped by tears.
I pray there are more, maybe 4, 6 or 10
And short be the parting, till we meet again.

Monday, May 22, 1995

I spoke to my family about the many positive changes in my life that I am excited about, especially my freedom from a mockery of a marriage, and my goals for furthering my education so I can accomplish what I have planned for myself in my new life alone.

Unfortunately, I had two marriages fail due to infidelity. I raised my two children alone on $250 per month total child support from their father. I had knocks, but I am hardy. I like my personality. I like my aspirations, dreams, and I look ahead to a wonderful future. Please give me a chance to be able to do that.

I promise when I get another head cold, I will never take another decongestant, antihistamine, antinauseant, or antidepressant again. I will use nasal spray, drink fluids, and get plenty of rest. I promise I will never push myself when I am physically ill again. Please let me go home, and let me take some time off to get some rest so I can become strong again.

If I must stay here at the Waterford Hospital, please put me somewhere that is smoke free as I am going to choke to death here coughing. I was also diagnosed with vasomotor rhinitis by a respirologist and allergist. This condition may be brought on by stress, hormones, etc and cigarette smoke. Since the early eighties, I have been very sensitive to cigarette smoke.

These writings release my pent up frustrations from being locked up in here, especially since I have certification exams to write on June 5th. If I must stay here, I want to go home and pick up my books that I have to study. I have invested $350.00 in this exam, so I would like to write it and pass.

After lunch, I went alone for another walk around the buildings here. While I was walking past the ball field, I stopped and watched a family from Kilbride having a fun game of softball. I was chatting with the Mom, who was sitting on the benches, enjoying the activities with her family. It made me feel good to see them, as a family, enjoying leisure time so effectively together.

After awhile, I left them and continued my walk and then met E., a patient at the Waterford Hospital for many years. I recognized him as one of the patients who had attended the movie last night. I asked him how he had enjoyed the movie and he said "it was literally horrible." I stated "that was why we had left as we thought the film was too frightening."

I asked "Who monitors the films that the general audiences of the Waterford Hospital watch?" I believe that movie, "Discovery" should never have been shown, because I feel it could be too disturbing to the patients.

E and I walked and chatted for a long time and then we sat on the grass and enjoyed the beautiful view from the top of the Waterford property, which overlooks the South Side Hills and Bowring Park. I could look to my right and see the windows of my locked ward. The windows reminded me of a prison and I had better get back there as it was approaching dinnertime.

I ate dinner and I knew I was beginning to feel better physically as my humor was returning and we were having a good laugh around the dinner table.

P., J., and I went for another walk after dinner and sat on the benches outside the main entrance and chatted until we were beginning to feel chilly as the temperature lowered while the evening progressed.

I was sleepy from all the fresh air, so I took a hot bath and went to bed around 2030. I slept until I was awakened for a bedtime snack. M. was standing outside the door, looking much better, but she said that she did not feel like eating or drinking anything now. I laughed and said "come along for the "social" entertainment." She laughed and we walked down the corridor together. It was good to see her laugh again. She is beginning to feel better.

Nurse's Progress notes
May 22/95 0800-1000

Patient continues to state she is in hospital a certified patient because of her second husband telling lies about her. She claims he is doing this because he is a 'psycho-path'. (I called him a pathological liar not a psycho-path)

She further states this can be confirmed by her father as he called her last night and told Judi that "all her problems were because of her second husband."

She again informed writer that she is not a manic depressive. She has "a lot of stressors in her life. She presented the argument of her diagnosis by saying "the doctors stopped my medications because they knew I did not need it." This confirmed to her that the medications were not necessary.

Patient continues to speak at length of the nursing profession. In conversation the patient is circumstantial plus, plus, however more in control with normal rate of speech.

The patient explained to writer the problems with nursing is related to the fact that nurses are "like crabs," they eat their young in order to survive and we should be more like dolphins as they circle around their leader and protect each other when needed.

She further states she will spread this to "all the nurses in the province of Newfoundland."

Unfortunately this analogy has been documented many times in professional nursing magazines and that is not my opinion alone. Many studies have been done on how older nurses are hard on each other especially the younger ones and why. During my university days, I wrote a researched paper named "Nursing Oppression or Nursing Profession"

Nurse's Progress notes
May 22, 1995 1000-2000 hours.

Patient is more subdued today, however placing numerous telephone calls to friends and family.

Informed writer she called Dr. Hughes (doctor at Health Science Centre), who is a good friend and explained her situation to him.

Also informed nursing student she was to sell her home in Newfoundland and move to New Brunswick.

She called a lawyer to begin divorce proceedings. (This is incorrect. I had called a lawyer to help me get discharged.)

Using off ward privileges.

She is keeping a low profile this afternoon. She was up to dining room for her meals.

She continues to state she does not need medications.

She requested to speak with doctor on call. Dr. Pratt visited and assessed this morning.

Nurse's Progress Notes
May 22, 1995 2000-2400

Judi has been active on the unit assisting other patients with their needs.

Tuesday, May 23, 1995

This is the second week of confinement and I have no idea how many more I will be locked up here. I awoke around three thirty and went to the desk to get some ice water and Kleenex.

I spoke to many people, including physicians today and was told that I still have to stay in here. When I looked on the large notice board near the desk, I noticed that my privileges were taken away from me again and I am not allowed to go anywhere or do anything. Why? I cannot believe that they are doing this to me. It is so unfair!

I had some visitors, including my son. I feel as normal as I have always felt, but I am not allowed to go home or even go for a walk.

A lawyer was in today to see me and she said she was going to phone Dr. Karagianis to tell him that I need another opinion from Dr. O'Loughlin. I told her that If Dr. O'Loughlin thinks I need Lithium then I will have to take it.

The lawyer told me she was familiar with Lithium She said that it was only a salt solution, and it is very safe. I wanted to learn more about Lithium, as the name itself struck me as being possibly dangerous. I asked myself "Don't they use Lithium in batteries?"

I requested the use of a Compendium of Pharmaceutical and Specialties (CPS) and began to read up on the drugs they wanted me to start taking. I knew that I was sensitive to most drugs and I did not want to take anything unless it was absolutely necessary, and I felt it was not.

The lawyer also said to me "you have no rights now, stay in your room, shut up, sit down and do whatever they tell you and take whatever drugs they give you or else you may never get out of here." That was a real reality check and I was in shock with that news. For the first time it hit me that I am really locked up in a psychiatric institution and I have absolutely no rights or freedoms as the Mental Health Act overrules the Human Rights Act. I really began to be scared and I do not scare easily, but I knew I had to very be careful what I said and did from now on.

I had a very boring evening as I could not go for a walk, so the time was very long. I kept to myself as I did not want to talk to anybody.

Nurse's progress Notes
May 23/95 0000 -0800

Awake at 0300 hours writing in her journal. Awake at 0500 hours for the day. She is anxious and worried about her mother's condition.

She wants discharge today.

Still expresses that she is not emotionally ill, but has only the flu.

States she has had a psychotic reaction to Trinalin. Labile mood.

0710 Judi was speaking with her brother at this time. She was upset with him because she is committed to the hospital and feels she may never get out of here.

She denies that she is manic depressive and didn't need any medications. She told him when she was officially diagnosed as manic-depressive, she would take Lithium.

She said her estranged second husband has convinced people she is crazy.

Sister-in-law called back and wanted to speak with Dr. Karagianis. She was informed that he was not available at this time and left a number that she could be reached today.

Brother states he will be working all day and his wife would be available to speak with the Doctor who is looking after her.

They re concerned that she may be discharged today and they feel she isn't well enough to go home. Mrs. Sparkes can be reached at (the number)

Clinical Clerk's Note
May 23, 1995

Patient was seen today, very unhappy about being hospitalized. She is still convinced that her present state was brought about by her upper respiratory tract infection and Trinalin use.

She feels well and wants to be discharged.

Patient has made significant improvement last week after being on Haldol for two days,

(I swear I did not take any), besides Haldol does not work that fast anyway. It takes weeks for that antipsychotic drug to be effective and reverse the symptoms of psychosis in a patient who has bipolar disorder.

Clinical Clerk's Note, cont'd

The patient insists she took none of her medications, but the record indicates she was indeed compliant at that time. When seen last week, she was organized, coherent and there was no evidence of active psychosis. The decision was made at that time to prescribe her medications on a whenever necessary basis and observe over the weekend.

It appears that the patient has deteriorated significantly since the medication was stopped. She was noted by the nurses to be interfering with the nurse's duties, asking for other patients to get their medications, etc.

She has also written a lot of letters,

(I wrote none, except my Appeal for Release, and in my journal)

She has made many long distant phone calls.

I phoned my father and the hospital to see how my mother was making out.

She showed "lability" of mood.

Just for a moment, consider my confinement and uncertainty of the forced medication therapy and think about what emotional state you would be in at this time.

Clinical clerk Note, cont'd

May 23, 1995

Also she showed initial and terminal insomnia.

Finally she wants to contact a lawyer and start proceedings against her unjust hospitalization.

She has also seemed to make decisions when not well- She says she wants to divorce her husband now, and as well she wants to sell her house, no matter how much money she will get for it. At this time she certainly seems to be a threat to her property.

When interviewed the patient was notably deteriorated since last week. She was not as well groomed, her hair was messy. She was very irritable when interrupted, and at times laughed inappropriately. She was almost tearful other times- i.e. her mood was quite labile, affect fairly congruent.

Speech was not pressured, but at time difficult to interrupt.

There was tangentiality (to change abruptly from the subject under consideration) **and a lot of circumstantiality,** (Giving full and precise details, incidental, related but not essential.)

No evidence of Delusions at this time, patient denies Hallucinations.

Patient is very adamant about not taking her medications, but it is in the best interest of the patient to receive same. She has been rapidly deteriorating and with no treatment she could again develop psychosis.

<u>Collateral history from co-worker reveals there have been a number of admissions at the Health Science Centre, under the care of Dr. Jain and the nurse "has no doubt about the diagnosis of manic depressive illness.</u>

After I obtained my records in 1996, and saw the above statement documented, I approached the nurse who I thought was responsible for this misinformation and she admitted to me that she "thought" that was my diagnosis and that she relayed this incorrect information on to the staff at the Waterford Hospital.

She was practically in shock and had to sit down when I told her that Dr. Jain had never diagnosed me with manic depression and I was never treated by him for that serious mental illness. She felt terribly sorry and offered me Dr. Jain's address and phone number in London, Ontario, so I could write him and get this problem of misdiagnosis corrected. It was a grave mistake on the nurse's behalf to impart that incorrect information to Dr. Karagianis, and I suffered terribly for it.

I wrote Dr. Jain, regarding this issue, but did not receive any replies. I phoned him when I was in London, Ontario in the fall of 1997, and spoke to him for a long time. I asked him if he ever thought that I suffered from bipolar disorder and he said "definitely not", even though during the medical malpractice lawsuit, he testified that he had diagnosed me with bipolar disorder in 1983, but did not write in on my chart in order to protect my job. What a stupid remark for a physician to make and the judge to believe! Physicians certainly stand together, even perjuring themselves to protect each other from medical malpractice lawsuits, to the detriment of patients everywhere.

I know now from being hospitalized with mentally ill patients, they really do not matter as they are people who do not seem to be a threat to anybody. They have been bullied so much by their care givers, upon whom they depend, that they are destroyed every way possible, and cannot speak up for themselves as they are controlled on psychotropic medication, possibly for the rest of their lives.

I saw a young patient with traces of antacid on her lips, complaining of a burning stomach, and the many others who were just helplessly wandering around and just needed some attention and understanding for another caring individual.

After I was diagnosed with a serious mental illness, I also realized that nobody took anything I said seriously again, as I was not considered credible, and definitely not worth concern, so who needed to be accountable to me? I could not be a threat to the establishment as there is really nobody to represent me and I could not represent myself while I was locked up and controlled by authority and drugs.

There is not much chance of the possibility of finding the medical or nursing establishment negligent, cruel or aloof when it comes to psychiatry because there are absolutely no safety nets that can keep patients from falling through the cracks, and no recourse once they have while their lives are ruined forever. Even though there are Mental Health Acts in place, they are being violated as was done in my case.

I was sure I was not mentally ill and I knew that I would be able to prove that some day. Perhaps by doing this, I would be able to give the supposed mentally ill a chance to become empowered. At least then, they could have a voice that would be considered credible, even in the courtroom, after they are emotionally, physically and sometimes even sexually abused by their professional care providers and others.

It is well known in the legal circles that any lawyer who takes on supposed mentally ill patients as clients, does so at his/her own risk as psychiatric patients are not considered credible.

I knew I was credible, therefore I spent many years, representing myself in a malpractice battle. I did this, not because I believed that I was going to win the lawsuit, or gain redress for the losses of my credibility, my career, my financial security, and my health, which were the result of negligence, but because I wanted to speak up and speak out about the abuse supposed mentally ill patients are exposed to by physicians and nurses behind the walls of hospitals, clinics and even in their own homes.

Clinical Clerk's Note
May 23, 1995

Patient is requesting a dinner pass with a friend. At this time, the patient is quite manic as described above, and it is against our judgment for the patient to leave the hospital at this time.

I spoke to son and daughter. Daughter feels patient is not well and has not been for awhile.

When in Fredericton, patient was noted by students to be very active and seemed like her thoughts were disorganized.

Over the phone daughter thought the patient was not well, talking very fast. I explained to daughter her mother's present management.

During that phone call, if my daughter actually imparted this information to the clinical clerk, she certainly added to the confusion by stating that inaccurate information, about the students noting that I was very active, and it seemed my thoughts were disorganized, when I was studying at University of NB.

During that second semester, the students voted me as their student/faculty advisor and asked me to take the 7th course, as I had already mentioned. My GPA average entitled me to the Cutler Nursing award at UNB for that year and I won a scholarship of $2500 to continue with my studies the following year.

Is that what happens when your thoughts are disorganized? I don't think so. When Karen spoke to me on the phone, while hospitalized against my will, perhaps I did talk faster, out of frustration of being locked up in a mental institution. Karen later testified during my medical malpractice lawsuit that she will always be sorry for not coming home when I was hospitalized until the death of my mother, her grandmother, which happened June 1, 1995, two weeks after my confinement.

By then it was too late for me and certainly too late for my mother, as I had already been forcefully drugged to the point of incapacitation with Haldol. My left hand is still abnormal and becomes spastic from taking these drugs against my will, or be injected.

After I reviewed my medical files in 1997, it was recorded on the Clinical Clerk's note Monday, May 23, that "son thinks mom is better" The staff did not take him seriously either.

Clinical Clerk's Note
May 23, 1995

Son thinks mom is better.... But that statement was ignored.

Nurse's Progress Notes
May 23, 1995 0800-2000

Judi has been tearful at times today. She remains over talkative. She is upset that she can't be discharged.

She is phoning everybody including the nursing director, lawyers and friends she feels will help her.

She is accepting she has to stay here fairly well. Her privileges have also been cancelled, which she finds a little restrictive. Speech is a little rapid.

Medications are ordered but Judi refuses them but indicated she may take them after she speaks with Dr. Karagianis in the morning.

Lawyer visited this evening and Judi says that the doctor has the right to treat her against her will. Judi is upset with this knowledge, feels this is barbarous, as she feels that she is now back to normal and blames her previous behavior on a reaction to Trinalin.

Says she needs to get an affidavit from nurses so that she wouldn't have to take treatment, but knows that this is very unlikely.

She is eating well, more settled this evening.

Her second ex-husband called, but she refused to speak with him as she blames him for her admission.

Nurse's progress Notes
May 23, 1995 2000-2400

Judi states she has been upset most of the day. She spoke with her doctor today and he told her that she could be forced to take her medication.

She denies that she has a problem that necessitates taking medications. She took the Haldol and Cogentin with much persuasion.

Question whether she spit the Haldol back in glass of water, but staff made her drink all the water.

She took Lithium also even though she stated earlier that she wasn't going to take it. She was settled to bed at 2330 hours.

By this time I had phoned the patient's complaints number that was written on the notice board several times and complained to the Patient's complaint's advocate. I had spoken to the nursing supervisor on many occasions as I personally knew her because of her marriage to one of my teachers in high school. I had also filed an Appeal for Release and nothing or nobody was helping me to get out of this frightening situation.

Some people think doctors can put scrambled eggs back into the shell./Canfield

The View of Bowring Park from the Waterford Hospital

Chapter 7

The Beginning of Forced Drug Therapy
May 24, 1995

Wednesday, May 24, 1995

They forced the drugs on me last night and I took them as otherwise they would have injected me. I slept well and woke up very drowsy. I can hardly open my eyes today. I took my bath and am waiting for my breakfast.

I feel so sleepy and drugged that I can barely hold my head up.

I ate my breakfast and went to the phone room to call my real-estate agent, regarding my house sale. Here they say it is irrational behavior, but it is financial survival. I also phoned the personnel department at work regarding my sick leave benefits. I have to get these benefits in place soon, as I need income.

I took the second dose of Haldol and Cogentin. I feel like this is the beginning of an unproductive life, living on social welfare. I will be reduced from being a very successful person to a welfare recipient by the staff of the Waterford Hospital. I have no choice but to go along with their regime as the mortals here have the authority to do this to human beings.

Two physicians have the right to ruin your life. I would rather be dead than to live like a drugged zombie, without any drive or goals in life. My zest for living will be destroyed. I will possibly never function normally again. The doctors believe I have manic depressive disorder, so I am forced to conform and take dangerous psychotropic drugs.

I began my therapy last night and today I can barely hold my head up, as I am so sensitive to these drugs. It is unbelievable what is happening and there is nothing I can do about it.

I had a meeting with Dr. Karagianis and they confirm that I must resume the drug regime and that I will be transferred in a few days to the Health Sciences Centre, where I have been employed for the past thirteen years.

My son Brian is on his way in to see me.

I have spent most of today in bed as the drugs they gave me are making me very sleepy. I experienced a very severe stomach pain that really hurt. I could not remain lying down as it was worse that way. After I took the Lithium it became worse, so I took some Diovol and warm milk and the pain finally went away. I went to sleep, woke up at three in the morning, and went to the bathroom. After that I slept until six.

I am still upset that I did not have any choice of the treatment and that I am being forced to do this. I still wish to speak to Dr. O'Loughlin and get another opinion. It is terrible when people cannot choose their treatments, especially when these drugs are so damaging to the human body.

Clinical Clerk's note:
May 24, 1995

Patient discussed at rounds. It is clear that the patient is quite intelligent and may seem to present herself well, but longer interviews and consensus from the staff show that the patient is not well.

She is quite manic off her medication and danger to her property.

A transfer is approved to Health Science's Centre as soon as a bed becomes available.

In view of the patient's compliance with her medications last night, she will get unrestricted privileges and accompanied pass this morning with her son.

Plan:
1. Transfer to Health Sciences Centre
2. Unrestricted privileges

Dr. Karagianis' first visit since admission:
May 24, 1995

(Illegible) given times 17 minutes. Discussed features of hypomania and need for medications.

Since she agreed to take medications last night she has already settled considerably. However she has no insight and insists that others have lied about her symptoms.

Agreeing to take medications, but does not think she is bipolar.

Nurse's progress notes
May 24, 1995 0800-1600

Judi is more settled since taking her medications, but feels tired and looks drowsy. She still feels that she doesn't need the medications and disagrees that she has features of bipolar illness.

She received visitors today, son and daughter and friends.

She was given back unrestricted privileges.

Health Sciences Centre have agreed to accept Judi in transfer when they have a bed and Judi is very pleased with this.

Clinical Clerk's Note:
May 24, 1995

Patient compliant with medications last night and is much settled since yesterday. She is much less agitated and irritable. She is convinced she does not need treatment however, and has a firm belief that her episode can be attributed to an adverse drug reaction.

She is feeling drowsy and tired and does not want to feel like it for the rest of her life. When seen the patient is slightly lethargic, but behaving appropriately.

Her mood is subjectively depressed, objectively euthymic (normal). Speech is normal rate and volume. Slight circumstantial, no tangentially. No insight.

Impression: Improving on Medications
Plan: As discussed in rounds in am.

Student nurse's note:
May 24, 1995 2000-2400

She spent a quiet evening in her room. She is complaining of drowsiness, blaming Haldol for same.

More approachable and less agitated than previous weekend.

She continues to have poor insight into her illness, stating that she has no reason to be here.

She informed writer that she was planning to appeal her case to the Review Board.

She complained of upset stomach, given Diovol at 2130 for same with no relief noted. She was given warm milk at 0015 hours and settled to bed.

Thursday, May 25th, 1995

It keeps bothering me how I came in here and did not have any say in my line of treatment. I was not explained anything. I was ordered to take drugs, without my consent. I was committed involuntarily without just cause.

I tried to fight the system, but that is impossible, because if you do not take the prescribed drugs, they believe they have the authority to hold you down and inject you.

This morning I went down to the chapel for a sing-a-long. I love singing as it always lifts my spirits. After lunch, I went to the hair dressers and had my hair cut. Around two thirty I went out on a pass with Mona Matthews. We went shopping and to Pizza Plus for supper, which I really enjoyed.

I arrived back to the ward at 1830, put on my lounging clothes and went to sleep. I was awakened by the nurses once for pills, but other than that I slept.

Clinical Clerk Note:
May 25, 1995

Patient complaining of drowsiness, stomach upset and low energy. She is very concerned about not being able to sell her house.

Objectively patient is euthymic, behavior appropriate. Well groomed. Affect congruent (state of agreement and harmony) Speech volume and rate normal. No tangentially and circumstantiality. No evidence of delusions or hallucinations. Insight Poor, judgment poor.

Impression:

Good response to Haldol, manic state settling

Plan:

Continue present management
Increase Lithium to 600mgms, by mouth, at night
Transfer to Health Sciences when bed is available.

Student Nurse's Progress Notes
May 25, 1995

Judi spent a relatively quiet day. She is still upset about having to take her medications. She believes it has an adverse effect on her. She is complaining of being drowsy.

She doesn't appear to be aware of the reason behind her admission. She reported that she is not ill, but merely has a "zest for life."

She has been ordered three or four accompanied passes whenever necessary. She has been using her unrestricted privileges well. No problems.

Nurse's Assistants Progress Notes:
May 25, 1995

Judi returned from pass at 1830 hours. She spent the remainder of the evening in bed. She is complaining of being drowsy.

Nurse's progress Notes:
May 25, 1995

She was approached with nighttime medications 600 milligrams of Lithium, Judi questioned same. She refused to take 600 milligrams but agreed to 300milligrams.

She says she had pain under her right rib cage for two days and is fearful her liver is giving out because of all the medications we are giving her.

She says she is not taking any more medications until her LFT's (liver function tests) are done. She says the doctors are trying to kill her and if they knew anything they would be monitoring her blood levels.

She was assured that she had blood work done since hospitalization, but she disagrees. Electrolytes and LFT's levels ordered for tomorrow morning.

Writer checked computer concerning blood work and TSH was 8.2, (normal is 3.5-5) but no copy of same on chart. Will pass on to doctor in morning.

It took ten days for the results of my thyroid tests to be available. My thyroid stimulating Hormone (TSH) was elevated, which indicates that the pituitary gland was working harder than normal to stimulate the thyroid gland to keep producing the thyroid hormones t3 and t4 that affect the body's ability to maintain itself with energy and other mechanisms. I was hypothyroid, not hypomanic.

I had been complaining of being tired since admission and instead I was diagnosed with mania, which is the opposite, and then I was forced to take Lithium, which indirectly slows down the thyroid gland even more. The fatigue I was experiencing is indescribable and very explainable.

Dr. Karagianis and then Dr. Craig ignored my thyroid malfunction, which was beginning to happen in 1995. They kept prescribing Lithium until I became seriously ill with fluid retention, swelling of extremities, shortness of breath, severe fatigue and muscle weakness.

After a year of suffering and getting worse, Dr. Amy Tong finally diagnosed hypothyroidism the summer of 1996 and then the Lithium was discontinued.

I have been taking medication for my thyroid gland ever since.[1] According to medical experts and research, my permanent thyroid malfunction was a direct result of Lithium as hypothyroidism and goiter are common complications of Lithium therapy. This problem can be usually managed by supplemental thyroid hormone.

[2] By November, 1995, my ECG was abnormal, (ST abnormality). I did not know this information until I obtained my medical records in 1997.

Back to my journal:

Friday, May 26th, 1995

I awoke at three-thirty in the morning and went to the bathroom. My throat was very dry so I proceeded to the desk to obtain some ice water. Immediately I went back to sleep again to awaken at six thirty.

I got out of bed feeling very sluggish, took a bath and got dressed, waiting for breakfast. I had to lie down again after I made my bed as I am so sleepy. It keeps dwelling on my mind how unfair this treatment is for me. When will they realize that I am not manic-depressive and let me get on with my life as I have always done?

---

[1] Dunner D.L., Meltzer, H.L. Fieve R.R. Clinical Correlates of the Lithium pump Am J Psychiatry, 1978; 135: 1062-1064.
[2] Joseph, Mary, M.D. and Vieweg, M.D., (1994/1995) "Electrocardiographic changes of sinus Bradycardia and Sinus Node Dysfunction among Patients with Therapeutic levels of Lithium." Depression 2: 226-231

It makes no sense why I have to take these drugs because I was perfectly fine for almost a week before they started the regime. "Where is this going to end?" I thought. I finished breakfast and had my blood taken to test for Lithium levels.

Lithium is one of the drugs used to treat manic depression. It is a very toxic drug that can cause[1] permanent neurological damage as there is a fine line between what is therapeutic and toxic.

While you are taking the drug you have to have your blood checked to ensure that your levels remain in the safety range, which for a normal person is around 0.8mm0/L to 1.2mmol/L. In my case, because I was suffering from undiagnosed fm/cfs/mcs, I was toxic well before the normal safety range.

Women are also more prone to Lithium neurotoxicity as they have less capacity to "bail" out Lithium and thus retain it intracellularly. (Dunner et al)[1]

In fact when a specialist, Dr. Eleanor Stein, a psychiatrist from Calgary, Alberta, reviewed my files, it was evident to her many times, according to my voiced complaints, that I had become lithium toxic at around 0.7 mmol/L, as I was experiencing symptoms of fatigue, weakness, drowsiness, nausea and vomiting and later blurred vision, ringing in my ears and loss of balance, plus cognitive difficulties, and later on a psychotic episode, when Prozac 20 mgms was prescribed with Lithium, which landed me back in hospital again at the Health Sciences centre in the fall of 1995 under the services of Dr. Craig.

[1]Apparently, it is possible that the "cellular' Lithium concentration increases, that would negatively affect the brain, and other organs are possible even though the serum Lithium concentration was within the therapeutic range. [1]Dunner et al)

[1]Some people (like me) are genetically predisposed to take up more Lithium intracellularly from a given dose than others. These individuals would be more prone to develop neurotoxicity. (Dunner et al) Specific risk factors for developing lithium neurotoxicity at therapeutic levels are rapid dosage regimes, concomitant administration of neuroleptics, *(Haldol),* pre-existing EEG abnormalities, genetic susceptibility, undetected cerebral pathology or pre-existing organic impairment. (Dunner et al)

In conclusion, any patient on Lithium who develops any symptoms suggestive of neurotoxicity regardless of other therapy should have their lithium discontinued, an EEG performed, and serum and red cell lithium measured, since irreversible syndromes can occur if Lithium therapy is continued. (Dunner et al)

Even though this medical information was available since the seventies, the good doctors who were treating me had little understanding about individual drug sensitivities, and completely ignored my symptoms of Lithium toxicity.

As misdiagnosis in itself is not considered negligence, the diagnosis of bipolar disorder was not really the issue. The negligence actually occurred because my physical symptoms of neurotoxicity were ignored for over a year and caused permanent injury to my body.

According to the statistics I researched in 2003 through[2] Canada's Adverse Drug Reaction Database, there were patients who have been injured like I was by taking Lithium. From what I can understand from the statistics I reviewed at the time, 50% recover, 25% have permanent neurological damage and 25% of those people die prematurely. I lived, but have to live with permanent neurological-endocrinological damage. (brain and thyroid)

Sinus node dysfunction of the heart can be induced by therapeutic levels of Lithium associated with hypothyroidism. Once Lithium is discontinued, this is reversible, but in some patients, as it was with me, it is permanent.

Even therapeutic blood levels of lithium can impair the central nervous system, both motor and sensory functions, and the cerebellum appears to be the central nervous system (CNS) region most

---

[1] Dunner D.L., Meltzer, H.L. Fieve R.R. Clinical Correlates of the Lithium pump Am J Psychiatry, 1978; 135: 1062-1064.
[2] Canadian Adverse Drug Reaction Database. http://www.cbc.news/adr/database/HealthDbservlet

often involved. If other drugs are added like Haloperidol, this further increases the risk of peripheral nerve damage.

I don't believe anyone should be forcefully given dangerous drugs like Lithium and Haldol that can cause permanent neurological damage, without the patient being in agreement with the diagnosis and treatment, especially when I was displaying symptoms of confusion, memory problems and fatigue, not mania when I was admitted.

I had a normal mental status since the second day of admission, May 15, 1995, and behaved appropriately except I lacked insight into being diagnosed with bipolar disorder, as that possibility had never been discussed with me, before that admission. Subsequently, I was assaulted with high dosages of drugs for over three weeks. This in my mind is utterly unacceptable medical treatment.

I have always had a medical history of sensitivity to medications and anesthetics. This has been documented many times on my medical charts. In May 27, 1992 when I was having surgery for endometriosis it was recorded on the nursing progress notes "complaining of numbness and 'pins and needles' feeling in right side of head and neck". She is complaining of "not feeling good." "She vomited four times since operation." This was again due to my sensitivities to drugs.

To find support for my accusations regarding neurotoxicity, I wrote the Lithium Information center, Dean Foundation, 2711 Allen Blvd. Middleton, Wisconsin, USA 53562 requesting information, and received dozens of articles regarding "Lithium Neurotoxicity", even with normal blood serum levels.

Listed below are some of the articles I received. After reading those articles, there was no doubt in my mind that my symptoms were related to lithium neurotoxicity.

The dates of publication of those articles were in line with the time when I was medically mismanaged. These articles were presented to the judge during my medical malpractice lawsuit.

1. **Bell, A.J. and Ferrier, I.N. (1994) "Lithium Induced Neurotoxicity at Therapeutic Levels- An Etiological Review." Lithium (1994) 5, 181-186. #24,427**

2. **S. Grignon, B. Bruguerolle (1996), "Cerebellar Lithium Toxicity; A Review of Recent Literature and Tentative Pathophysiology." Therapie, 1996; 101-106, #25,862**

3. **Verdoux, H & Bourgeois, M.L. (1990), "A Case of Lithium Neurotoxicity with Irreversible Cerebellar Syndrome. Journal of Nervous and Mental Disease, 178, 761-762**

4. **Dorus, E., Pandey, G.N., & Davis, J.M. (1975) Genetic determinant of lithium ion distribution. Archives of General Psychiatry, 32, 1097-1102**

5. **J Jefferson, James W., and Griest, John H., (1994) "Lithium in Psychiatry, A Review, CNS Drugs 1(6) p448-464, #4724**

6. **Jefferson, James W. and Griest, John H. (1979) "The Cardiovascular Effects and Toxicity of Lithium, Reprinted from "Psychopharmacology Update: New and Neglected Areas." by Grune and Stratton, Inc.**

7. **Tetsuya Numata, Haruhiko, Abe, Takeshi Terao, and Yasuhide Nakashima, (1999) "Possible Involvement of Hypothyroidism as a Cause of Lithium-induced Sinus Node Dysfunction." PACE, vol. 22: 954- 957, #29,335**

8. **Joseph, Mary, M.D. and Vieweg, M.D., (1994/1995) "Electrocardiographic changes of sinus Bradycardia and Sinus Node Dysfunction among Patients with Therapeutic levels of Lithium." Depression 2: 226-231**

Back to my Journal: Friday, May 26, 1995

I tried to contact my real-estate agent, and my lawyer again regarding the sale or rental of my house, which was now vacant, as the tenants had moved out the previous month. This admission to the hospital is going to ruin me financially as well, unless I sell the house soon, or get another tenant from Hibernia, one who pays higher than average rent.

Mona came to get me at eleven o'clock to go out for the day. The nurse made sure I took my noon dose of Haldol, before I left and gave it to me an hour early. The drug makes me sleepy, so I must fight off that feeling.

Haldol is another dangerous antipsychotic drug that also may cause serious neurological damage, like twitching and muscle spasms. In fact, to reduce those serious side effects, they give you another toxic drug, Cogentin to prevent those extrapyramidal, (EPS) (other than the pyramidal tract) symptoms. I am being forced to take all those dangerous drugs against my will, especially since I have had a perfectly normal mental status.

Mona and I went to our friend Betty's apartment for lunch. While we were out, Mona offered to loan me $1,000, which I accepted and we went to transfer that money into my account that was badly in need of it. Mona said she wanted to help me out now as she appreciated so much all the things I had done for her before, like recommending her to do home care for my elderly friends who needed it. They could not have gotten better care as Mona was the best.

"That's what friends are for," I thought. I will definitely pay her back in August; before she goes on her holidays as hopefully I will be in a better financial position by then.

I arrived back to the Waterford Hospital at 1600 and took another nap as soon as I reached my bedroom. I ate well during dinner, and then Ford, Joy and my cousin Susan visited me in the evening.

I am still very tired and sluggish. I took a bath, and after I received my drugs and went to bed, my stomach began to burn. I drank several ounces of Diovol and two Rolaids, but my stomach pained so much, I broke out into a sweat. I finally got to sleep.

My stomach was burning from the effects of the liquid Haldol. Antacids are given in large amounts to counteract this effect, which really increases the calcium level in your body to disrupt the normal blood chemistry.

Clinical Clerk's Note
May 26, 1995

Patient is again complaining of drowsiness and insisting on her medication being decreased. No abdominal pain since two days ago.

Objectively, she is no different today. No evidence of formal thought disorders, no delusions.

Impression:

Improving manic state.

Plan:

Three to four hour accompanied passes, prn (whenever necessary)

Dr. Karagianis' note
May 26, 1995

Unable to meet with her today. She is on a pass

Nursing Progress Notes
May 26, 1995 0800-1600

Judith was up and around the unit throughout the morning, compliant with meals and medication, pleasant and cooperative on approach.

She left on a day pass at 1115 hours, had her noon medications before she left and is due to return to unit at 1600 hours.

She also had accompanied day pass over the weekend.

1600-2400

Judith had good evening, spending her time in bedroom, had visitor in evening. She ate lunch and taking medications.

Saturday, May 27th, 1995

I awoke in the morning feeling very miserable. I took a bath, got dressed, but I have no energy to do anything.

I took my medication during breakfast and spent all morning lying on my bed listening to music. I went to the cafeteria for lunch.

I also convinced the nurse not to give me the noon dose of medication, as I was too sluggish.

I have been lying down all afternoon, but I had the courage to get up and go for a walk around the building as it was cool outside and that always made me feel better. Afterwards I did not feel quite as drugged, and I have a little more energy than I had this morning.

I hope I can convince the nurse at dinner time that I do not need the drugs. I do not want to have to take them tomorrow, as I will not enjoy my trip with Mona to Sibley's Cove to see my parents.

The nurse suggested that I take my dinner time dose of drugs with my bedtime ones so I may have a more enjoyable evening by not feeling so tired. I took a hot bath and Ford and Joy visited for awhile. I took my medications, and settled down for a good night's sleep.

Nurse's Progress Notes
May 27, 1995 0000-0800

0030 Judi came to the nursing station complaining of not feeling well. She complained of moderate to severe epigastric pain. She was given Mylanta 30cc at that time and vital signs were taken.

Blood pressure was 116/78, pulse 88, respirations 20, temperature 36.6.

She did get relief after Mylanta (Diovol) was given. Dr. Gross notified of Judi's condition.

I checked on her again at 0115 and pain had just about resolved and Judi settled to sleep.

Nurses Progress Notes
May 27, 1995 0800-2000

Judi has been napping periodically during day. She complained of feeling drowsy and lethargic on Haldol.

She says she feels as if she just had an anesthetic and she can't function. She says the left side of her neck and face are numb. She blames this on Haldol and refused to take it at 1200 hours.

She says she will take this dose at 2200 hours tonight as it is only ordered three times a day.

She states she is so drowsy that she is unable to go on her pass.

She feels she is being treated unjustly and against her will.

She still maintains that she had a reaction to Trinalin.

She says that she has been asked to resign her position, but has refused. (I don't remember who phoned me and asked me to resign my position. I had forgotten that information until I had my memory jogged when I obtained and reviewed my hospital records in 1997.)

She is worried over finances. She doesn't have any sick leave, so she is ineligible for any pay if she doesn't work.

She says this may force her into selling her home.

She still does not accept a diagnosis of bipolar illness, so she is not agreeable to an increase in Lithium dosage.

Nurse's Progress Notes.
May 27, 1995 2000-2400

Judi spent a quiet evening. Lunch was taken. She spent most of the evening in her bedroom.

Sunday, May 28, 1995.

I awoke at seven and took my morning bath before breakfast. I was ready to go out at nine and Mona arrived ten minutes later. Mona and I had a pleasant drive to Sibley's Cove, except that we had to stop twice for me to vomit, as I was getting motion sick while riding in the car.

We arrived around eleven thirty and Mom was so glad to see me. She said to Mona, "You have brought me the best gift ever, my daughter." I hugged my mother for awhile and with tears in my eyes, I sat down on the floor in front of her lounge chair where she was seated.

I gave her a bag full of clothing that I had brought for her. Some of the clothes had belonged to Mildred, and her sister Betty had given them to me for Mom. I had also bought some items at the clothing store at the Waterford Hospital Used Clothing Store. My mother was so pleased to receive them as most of the clothes were new and were exactly her size.

This was another example of the misunderstanding about spending large amounts of money, buying my mother clothes that she did not need.

We talked and talked until lunch was ready. She seemed so contented and at peace with herself. She kept telling me that she wanted to say a prayer for me so that I could survive my days in the Waterford Hospital. She said she knew my future would be blessed, and for me not to worry as she understood God's plan for me. She also said she knew why my heartaches cut so deep, to give me strength to persevere and wake his nurses from their sleep.

There was such a rhythm to her words and rationale to what she was saying that I found a notepad and pen and wrote down her words, and also kept them in my heart.

She told me again, as she had told me many times before, that when she could not sleep, her strategy she used to help her was to repeat "please clear my brain and help me sleep."

It had been very difficult for me to remain asleep for long periods of time when I was first confined against my will in the Waterford Hospital. It was torture! I often would wake up frightened and in need of comfort and security.

My mother's prayer helped me.

I shall lay me down,
In peace and sleep,
For I know my God will shield me from all harm.

Now you awake and see the light,
God has helped you through the night,
Prepared your soul for nobler things,
And gave you wings for greater flight.

I understand God's plan for you,
And why your sorrows cut so deep,
To give you strength to persevere,
To wake his nurses from their sleep.

Joanne, Mona's daughter, who was also my mother's caregiver, had a beautiful salmon dinner prepared for us. After dinner we watched a parade through Sibley's Cove with cadets, lodge people, and the war veterans.

The gardens around the old homestead looked lovely, with well groomed lawns, neatly pruned shrubs, and a variety of flowers. Dad does most of the gardening now. He must be spending a lot of his time there making everything as beautiful as possible for my mother to enjoy. He has moved most of her favorite plants to both sides of her wheelchair ramp so that she can see them through the living room window when she is sitting inside.

My brother Jim, his partner Lang, and his son David, also visited Mom and Dad while we were there, and left shortly before we did. I spent a pleasant afternoon with them all, but we left to drive back to the hospital around 3:30 PM, as my curfew was coming soon.

I got motion sick again on the way back to St. John's. We arrived around 6:30 PM, too late for dinner, but the orderly got me a tray of food, which I appreciated and enjoyed very much as my stomach was empty.

(I feel much better tonight. I am not going to take my drugs until bedtime. This is one of the best days that I have had since I have been confined here.)

I took my bedtime medications and went straight to bed. I slept very well throughout the night, with the exception of going to the bathroom twice. I believe the Lithium is doing that. I woke up with my bladder full, very dehydrated, and craving oranges to quench my thirst.

Nurse's Progress Notes
May 28, 1995 0800-2000

Judi remains more settled, although doesn't like taking Haldol because of the fatiguing side effects.

She seems more agreeable to take the increase in Lithium if the Haldol were decreased, but sees no reason to take any medication.

She has gone on a day pass with her friend and plans to visit her parents in Sibley's Cove.

1800

She returned from day pass, states things went very well. She is in good spirits; mother is out of hospital and feeling much better.

Judi was very pleased with the visit.

She didn't take 1800 hour medications. She will take them at 2200 hours tonight.

2000-2400

Judi took medications at 2200 hours with no problem although still took only 300 milligrams of Lithium, instead of the prescribed 600 milligrams.

She finds the Haldol liquid causes moderate to severe gastric distress. She was given Diovol with nighttime medications. She was settled to bed

Monday, May 29th, 1995

I woke up at seven in the morning and got up right away. I took a bath and got dressed. After breakfast, my friend Lee called and informed me that she was working in the clothing store, and that *we could go out for lunch together. I went to the clothing store to see her, and bought Dad two pairs of pants that I thought he would like, at $5.00 a pair.*

The time is so long here as there is nothing for me to do. I changed my bed, cleaned up my room and have nothing else to do now until lunchtime. I cannot go into the TV room, as there is too much smoking going on there, and the other TV room is locked, because it is booked for a special group therapy meeting.

I came back to my room and tried to sleep, but could not. It is so boring here. It is unbelievable! The time is endless. (I also spent time earlier in the games room playing hockey and Chinese checkers. I spoke to Dr. Karagianis for awhile.)

Lee and I had a lovely lunch in the cafeteria, which I enjoyed. After lunch, I went for a walk.

I received a huge fruit basket from the staff at the Health Science O.R. It made me cry for a second. I ate some fruit and chocolate, felt nauseated and vomited up my lunch. I did not eat very much for supper, as I still felt ill.

After supper, I slept most of the evening. Then I took a bath, dressed for bed and waited for my drugs before I retired for the night. After I took the medication, I slept off and on until seven thirty in the morning.

Clinical Clerk's Notes
May 29, 1995

Patient had a fair weekend, enjoyed her passes. She still feels like there is nothing wrong. She wants to be discharged.

She was complaining of abdominal pain in the evening and feeling very lethargic. She attributes this to Haldol and wants the dose to be decreased.

She refuses to take 600 milligrams, only taking 300 milligrams. Again she feels there is no need for it and thinks her levels are fine with the low dosage (last Lithium level was down to 0.46).

She wants to "make deals" – suggests she will take all the Lithium if Haldol is decreased.

She is also saying if Dr. Jain tells her she is bipolar she will believe him and may take her medications.

At this point however, she insists that all her symptoms were caused by a drug reaction and thinks all relapses can be prevented by not taking decongestants, not by taking Lithium.

Today patient is well groomed and dressed appears drowsy. She behaves appropriately. Mood subjectively depressed, objectively euthymic. There is no evidence of formal thought disorder, No tangentiality, no circumstantiality, Speech rate and volume normal Not suicidal, no (illegible)

Impression:
No (illegible) mania when medicated.

Plan:
Continue present management.

Dr. Karagianis' second assessment.
May 29, 1995

She is considerably settled. She is requesting decrease Haldol-complaining of drowsiness. She is only taking half of prescribed Lithium.

Still unencumbered by any insight.

Today she is not euphoric. There is no elation, no pressure of speech.
Plan: Refer to Dr. Coovadia for assessment for Renewal of Certification.

Nurse's progress Notes
0800-2000

Judi was seen and assessed by Dr. Karagianis. She requested a decrease in Haldol. Haldol was decreased to 4 milligrams, by mouth, twice a day and at night.
Judi continues to feel that medication is not necessary.
She is up and about the unit today
Her pressure of speech is improved.
She is consulted to Dr. Coovadia who will see her tomorrow concerning certifiability. She has very poor insight into illness.

Clinical Clerk's Note
written on the consultation form to Dr. Coovadia
May 29, 1995 1317

This 49 year old female was admitted two weeks ago. She was agitated, had gross evidence of formal thought disorder with loosening of association, tangential thoughts and incoherence, acutely disorganized.
She was certified upon admission.
She has since improved on Haloperidol 5mg t.i.d, but she continues to have no insight and there is concern about her compliance once discharged. She is only taking half of her Lithium.

Nurse's progress Notes
May 29, 1995 2000-2400

She had a bath this evening and continues to complain of sedative effects of medications and looks sedated this evening. Haldol was decreased today.
Judi still has little insight into her illness and rationalizes events prior to and following admission.

The best doctors in the world are Dr. Diet, Dr. Quiet and Dr. Merryman. / Swift

Mom's wheelchair walkway that Dad had build for her

Chapter 8

Recertification under Mental health Act
May 30, 1995

Tuesday, May 30, 1995

I awoke, took my bath, and washed my hair before breakfast. Since breakfast, I have been lying on the bed listening to music.

The President of Newfoundland and Labrador Operating Room Nurses' Association, just phoned me to tell me that I have been awarded the Operating Room Nurses' Association of Canada Scholarship, (ORNAC), worth $1,500 for next year's studies.

I began to cry again, as right now, with the drugs I am taking, I cannot concentrate or comprehend enough to read anything. I can't study for my operating room certification exams. How will I ever function normally again if I have to keep taking these drugs? I am finding it difficult to even write. My hand gets tired and my writing is beginning to look very bad, like scribbles.

I spent most of the day lying in bed and feeling awful. I ate dinner at five, even though I knew I was going out to dinner at Greg Johnson's at eight in the evening to sign a rental agreement for my home.

I know I am gaining weight, so I have to cut back on what I am eating. I will be like a blimp by the time I leave here, if I keep eating like I am.

Judy picked me up at seven thirty, and drove me over to the Johnson's. I met the family who are going to rent my house. They want the house furnished, so I have no other choice now but to vacate Brian's apartment by the twentieth of June, and move my furniture from there back to my home.

Brian and I had sub leased my furnished apartment in his basement to one of the hockey players, on the St. John's Maple Leafs, while I was in University in New Brunswick, but now he has gone back to Montreal for the summer. I was going to live there until I return to university in September, when he would return again. But I am locked up in the Waterford Hospital with no idea when I will be released. I have lost complete control of my life!

I will find a furnished room to live more cheaply, and hopefully Brian will be able to find a new tenant. This will help him pay his mortgage, as he is only working at a low wage while he is studying towards a business degree at Memorial University.

It looks like I will be able to lease my house to an employee of Hibernia. Those engineers pay well, so at least we will not lose our homes. My God, I never had to juggle finances like this to survive, but I am determined to do anything that is moral and legal to keep afloat. Dear God, I have to get out of here! What a mess this confinement is making of my life and financial situation! It is now going to negatively impact Brian's financial security as well.

I arrived back to the Waterford Hospital and I began to cry, but stopped immediately as crying was going to do me no good. I will get myself out of this financial mess once the house is rented, and I can get back to work, but I have to get out of here first, so I have to be strong.

I remembered what my lawyer had advised me when she visited me, before I started taking the drugs, so I had to remain calm and collected while being controlled. It wasn't easy but I had no other choice now.

Clinical Clerk's Notes
May 30, 1995

Patient is still complaining of drowsiness. She says she spends twenty hours per day in bed. She says she cannot read, exercise, because she gets too tired.

She has no abdominal pain. She has no stiffness or other extrapyramidal effects, (eg. Tremors, contractures, etc. side effects of Haldol)

She took all Lithium last night.

She was seen by Dr. Coovadia, certification signed.

Dr. Coovadia wrote the following on the Certification of Authorization by Physicians for Involuntary Admission and dated it May 30, 1995

Problem:

Recent history of grossly disturbed behavior as evidenced by motoric hyperactivity, accompanied by mood congruent delusions and formal thought disorder.

Past History

Similar maniform episode ten years ago, together with a history of sustained depression. She has a positive family history of mood disorder.

The maniform episode I had ten years prior was diagnosed as being "drug induced". I had no history of sustained depression and there was not a positive family history of mood disorder. I had a normal mental status now since May 17, 1995, except I had no insight into having bipolar disorder.

I had recovered from drug induced psychosis as soon as the medications that caused the psychosis were out of my system. At the time of this admission, I was in a confused and psychotic state. The cause of the psychosis was unknown. I had never been diagnosed with bipolar disorder up to that time. There were three differential diagnoses documented, one being "substance Induced Mood disorder."

Current Mental Status Exam:

She has responded well to treatment to the point that she is no longer motorically or cognitively hyperactive. She however continues to harbour her delusions, but with less intensity.

What is of concern is that she has no insight into her condition and of the potential serious consequences thereof.

In light of the above, she is likely not to comply with her medications and runs a risk of relapse.

She remains in need of medications.

Recommendations: Certify.
Dr. Coovadia, May 30, 1995.

Dr. Karagianis had consulted Dr. Coovadia to examine me for re-certification under the Mental Health Act of Newfoundland and Labrador.[1] That Act clearly states that, "in the interest of safety, safety to others or safety to property, may without his or her consent be admitted to and detained within and treated at a treatment facility." I was in no danger to anybody and had a normal mental status, except I did not think that I had bipolar disorder so I displayed no insight.

---

[1] RSN1990 Chapter M-9 Mental Health Act of Newfoundland and Labrador. Admission to treatment facility 1971 No8O s5 (1)

I remember every incident that occurred when I was confined against my will because I had a normal mental status, and had documented what happened every day, up to and including the recertification on May 30.

I have no recollection of Dr. Coovadia interviewing me for the purpose of recertification, and I had not written that important detail in my journal.

[1](a) The Act also states that "the physician has made personal examination of the patient." There is no evidence on my chart that Dr. Coovadia had actually seen or interviewed me for recertification purposes. The nurse testified in Discovery that she saw Dr. Coovadia going down the corridor, with his staff, but could not recall him entering my room.

[1](b) "The physician is satisfied that the patient continues to suffer from a mental disorder to the degree specified in subsection 5 (1)." Dr. Coovadia also wrote on his certification form that I "was no longer motorically hyperactive or cognitively hyperactive." I believe that those recertification papers were completed May 30, 1995, in violation of the <u>Mental Health Act</u>.

Upon my lawyer's advice, I filed a complaint to the Newfoundland and Labrador Medical Board regarding Dr. Coovadia and his recertification without examination. Dr. Coovadia wrote an official letter to Dr. Robert Young, President, and stated that he remembered talking to me, as I[1] "had complained of mild and vague abdominal pain, which was difficult to localize on history." He continued to write[1] "I elected not to physically examine Ms. Day, but stated that I would pass on her complaint to (the clinical clerk), which I did." That information was not accurate as I had complained only one night, May 26, of severe epigastric pain that resolved itself early in the morning of May 27. After that day, I did not have any problems with stomach pain. The Clinical Clerk's Progress Notes of May 30th also state "She has no abdominal pain."

Dr. Coovadia also stated in his letter to the medical board[1] "Her only 'psychiatric' complaint revolved around being illegally detained in hospital." Nowhere in my journals did I write that I was illegally detained, as I was not aware of that fact until I reviewed my charts in 1997. At that time, I studied the <u>Mental Health Act</u> for the first time, and came to the conclusion that there were violations of that Act and that I had been illegally detained.

During the stay at the Waterford Hospital, I was insisting that I wanted discharge, but my lawyer explained the legalities of my admission to me. I abided by them, thinking they were well founded at the time. I spoke of "violation" of the Mental Health Act for the first time during the Discovery on February 23, 2000, and amended my Statement of Claim to include that issue at that time.

Dr. Coovadia also wrote in his response to the Newfoundland Medical Board that[1] "speech was not pressured, mood was not elated and she was not objectively depressed." I had never been "elated." During those five weeks being detained against my will, I was far from being "elated", which is a symptom of mania.

There is no record on my chart of that observation of elation. I was physically ill with an upper respiratory infection. My mother was dying in hospital. I was very upset that I could not get released from hospital, and forced on drugs that I knew I did not require. I was definitely not elated! If I were "brittle and guarded," it was a normal response to the cruelty and barbaric treatment that I had been receiving for two weeks by then.

Dr. Coovadia also stated in his letter to the Medical Board that[2] "it was extremely difficult to obtain any spontaneous information from Ms. Day" when he interviewed me. I gave spontaneous information to everybody else who would listen, that statement was definitely inaccurate.

There was not one shred of evidence on Dr. Coovadia's consultation report that he had actually interviewed me, and no staff member could testify that he actually did.

---

[1] RSN1990 Chapter M-9 Mental Health Act of Newfoundland and Labrador. Certificate of Renewal 1971 No80 s8 (1), (a) (b)
[2] Dr. M. Coovadia's Letter. Confidential-Newfoundland Medical Board only. Sent to Legal Council August 01, 1997.

Dr. Coovadia had moved out of the country some time after 1995. During the trial, the judge asked that Dr. Coovadia be brought back to Newfoundland to testify regarding his examination of me before he signed the recertification papers, but that never materialized.

My testimony of the events was not convincing enough for the judge to rule in my favor, so he decided that the recertification was indeed valid, even though I am positively sure Dr. Coovadia did not personally examine or interview me. If he had, he may not have signed the form to detain me against my will, for another four weeks of torture, as I had a normal mental status, except that I lacked insight into my illness.

The Appeal for Release Form that I had completed on Sunday morning May 22, 1995 had never been acted upon either, so I was also denied, by the hospital administration, my right to appeal the involuntary confinement. Dr. Coovadia had also written in that infamous letter (1) "I recall that after recommending her continued certification, I ensured that she was aware of her right to appeal." Really?

Clinical Clerk's Notes, cont'd
May 30, 1995.

Patient well groomed and dressed. Her behavior is appropriate, spontaneous conversation, but no spontaneous smile.
Mood subjectively and objective euthymic, affect depressed.
Speech rate and volume are normal, no evidence of thought disorder.
No hallucinations, denies previous delusions.
No suicidal or homicidal thoughts.
Still no insight!
Impression: controlled on medications.
Plan: Discuss tomorrow at rounds.

Nurse's Progress Notes:
May 30, 1995 0800-2000

Patient is complaining that Haldol is making her very tired.
She spent day sleeping in bed and talking on the phone.
When approached by writer, patient stated that she was too tired to talk.
She requested medical certificate for employment purposes.
She was recertified today.
Presently sitting in lounge watching T.V.

2000-2400

Returned from pass at 2300 hours, said same went well.
She took nighttime medications, and then settled to bed.

Wednesday, May 31, 1995
Despite my frustrations, I had a good night's sleep and only awoke twice during the night to go to the bathroom. I got up at seven, took a bath, washed my hair, and dressed and ate breakfast. I did my laundry and had another boring day to look forward to.

Lunch and dinner were as usual. I tried to eat less as I feel I am gaining weight since I have been in here. I hate that idea, as it only means that I will have more weight to get rid of when I get out of here, if I ever do.

I am trying to find a place to live for July and August, as I have to move out of Brian's apartment, as the furniture will have been moved back into my home at Inglis Place by then as the house is being rented furnished.

This will bring in more income, but it means more juggling with my belongings, as I have to do what needs to be done for financial survival. And I am supposed to be mentally ill? Well, if I am crazy, I know for sure that I am still responsible for my survival and that I am not stupid, according to the scholarships I have received this year to do my Masters in Counseling.

Clinical Clerk's Notes
May 31, 1995

Patient discussed at rounds. Since regular medications have been started the manic episode has settled considerably.

She is not aggressive, sleeps better and speech has slowed as well as there being no tangentiality/circumstantiality.

However, she complains of being drowsy and tiring easily and insists on Haloperidol being decreased again.

Last Lithium level was 0.33 and since patient has been compliant with the full dose for the past two nights, levels will be checked again Thursday.

If the levels are low, Lithium will be increased to 900 milligrams and Haldol decreased.

As well, patient asked for a form to be signed to make her unable to work until the end of August.

This is not advisable and patient will be reassessed weekly re return to work.

Plan:

1. Check Lithium levels, and then adjust accordingly.

2. Reassess Haldol dose Thursday.

   Discussed management plan with patient, as discussed above. She seems agreeable.
   She is still complaining of drowsiness and asking for Haloperidol to be decreased.
   Form filled out re: disability to work.
   She is still waiting for transfer to Health Science Centre.
   She is cooperative, compliant with medications.
   No evidence formal thought disorder.

Impression: Settled on medications.
Plan: as discussed on rounds.

Nurse's Progress Notes
May 31, 1995 0800-1900

Patient spent most of day in bed, except for meals.
She is complaining of Haldol making her drowsy, inquiring about
getting Haldol reduced.
She is asking about date of discharge.
No other complaints or problems.

2000-2400

Judi spent a good evening. She took nighttime medications with no problem. Mood is good. She is settled.

I had no idea that I had been recertified as I thought I was still waiting to be transferred to Health Science Centre. My judgment was never compromised as indicated by my hand written journal. I signed employment insurance papers, appeal for release papers, and rental agreement for my home. Still I was recertified to forced confinement and treatment for bipolar disorder, the second most serious mental illness next to schizophrenia, a diagnosis that had never been discussed with me up to that time in my life.

No man is more worthy of esteem than a physician who exercises his art with caution and gives equal attention to the rich and poor./Voltaire

Climbing Mountains, Mt. Carleton, New Brunswick

Chapter 9

My Mother's Death
June 1, 1995

Thursday, June 1, 1995

My sister-in-law Joy came into my room in the morning and as soon as I saw her face, I knew something was terribly wrong and I uttered, "Mom is dead." "Isn't she?" "That is right?" "Isn't it?" "I knew this mess I am in was going to kill her." "She's dead!" "Right?"

She did not have to answer. I knew it was "yes." She proceeded to tell me how she had died peacefully, at home, like she always had wanted to do. She died, after eating her breakfast, while lying on her bed.

My elderly mother was definitely over sedated as she was not used to taking sedatives or sleeping pills as she was as sensitive to drugs as I am. I know she was sedated because the nurse had told me that when I spoke with her over the phone.

According to CPS, <u>The Canadian Drug Reference for Health Professionals</u>, Ativan is to be used cautiously in elderly patients, and patients with limited pulmonary reserve because of the possibility that under ventilation and/or being hypoxic, cardiac arrest may occur. She died suddenly of a cardiac arrest, just after finishing her breakfast and taking her medication.

I had always thought it was the Ativan that killed my mother until Dr. Nell disclosed during the malpractice trial that he was on leave of absence while another doctor was relieving him, and Mom had been sent home on Lasix 40 mg daily, without monitoring her electrolytes. That also contributed to her untimely sudden death. This was another death caused by drug overdoses and interactions added to the thousands who succumb annually from these serious issues.

A wave of peacefulness came over me. I knew the last time I had seen my mother alive that she realized it was the last time that we would ever chat again. She had told me everything that I needed to hear from her for the rest of my life. She knew exactly what lay ahead for me and that I would be blessed and content, fulfilling the purpose that God had in his plans for me. I had to suffer. I had to become strong. I had to fulfill that purpose whatever it would take.

I believe every ounce of strength and energy she could spare from her body, she allowed it to flow into mine, which helped me to become even more courageous, and strong enough to be able to cope with this unjust and cruel confinement with forced drug therapy that was making me so unwell.

I believe my mother knew that I was strong enough now to live my life without any more of her advice and help. Despite all the suffering throughout her lifetime with cancer, diabetes, and strokes, she always remained positive and was contented with her lot. She continuously spread positive thoughts to others. I was so blessed to be given such a wonderful mother for almost fifty years.

Since my mother's death, I have always felt her presence, which gives me so much peace of mind enabling me to cope with whatever trials and tribulations I have to face. I can live my life now and meet its challenges with the utmost courage, contentment, and control.

I want to share the beautiful poem that my mother wrote for me during the time of the breakup of my second marriage, when I was so devastated that my second marriage had to be ended.

I have always tormented my siblings when we got together that I had my own poem written about me. I guess it made up for being the middle child in the family, with two baby brothers who came along within eleven months when I was five and six years old. Can you imagine the displacement that I felt, even though we had the most loving mother, who would be every child's dream. Sometimes there was just not enough of her to go around, so I became her little helper. In this role I always felt loved and needed.

To my Daughter

My mind slips back to forty five,
A little angel came my way.
With eyes so bright and hair like flax
Sent down from Heaven, she came to stay

When I beheld my lovely child,
How my heart filled with love and joy.
I had three, thought life complete,
Two pretty girls, one lovely boy.

In childhood and your teenage years,
You were so precious, kind and good.
You helped me with those menial tasks,
Doing with love the things you could.

Your adult years were hard and sad,
And many a lonely hour you spent.
Your children both a blessing were,
They filled the void and brought content.

As you matured, your life was full.
You are revered and loved by all.
When'er you found a friend in need,
How quick you answered every call.

I'm sure your future will be blessed,
As you have sown, so shall you reap.
The many times you've stooped to help,
God in his records, He shall keep.

I left the hospital with Joy, and drove to Sibley's cove, so drugged that I could hardly hold my head up. The wake and funeral are such a fog in my mind as I was so incoherent from drugs, but I still managed to stand at the altar with my brother Ford, facing the congregation, as we took turns reading the verses of this poem that my mother had written and was one of her favorites.

'Tis Springtime

'Tis Springtime and the snow is melting,
Little brooklets everywhere.
Now has passed the cold, cold winter,
With its skies so grey and drear.

'Tis Springtime and the breeze is softer,
Warming up the springtime air.

And the soil is getting warmer
For the bulbs now sleeping there.

And now from underneath their blanket
The bulbs and seeds awake from sleep,
And when they're touched by God's warm sunlight,
The leaves through earth will slowly creep.

'Tis springtime with its longer twilights.
The crimson sun sinks in the west.
The busy world, man, beast and flower,
Is gently lulled to peaceful rest.

'Tis Springtime with the birds returning
God's feathered choir that sing so sweet
They fill the whole wide world with music
Then how can we, mere mortals sleep?

God taught each bird to call its mate,
And they will come from far and near.
The bluebird, robin and the sparrow,
They know each call so sweet and clear.

'Tis Springtime and the birds are busy,
Building nests and making love.
And soon you'll hear the babies chirping,
As parents guard from boughs above.

'Tis Springtime and on Easter morning
God's own Son from death arose.
And walked again within the garden,
Where the Easter Lily grows.

Each Springtime God renews the promise,
That has been made for you and me.
That we'll inherit life eternal
For Christ arose and so shall we.

    The one thing that I will never forget though was seeing my mother resting in her coffin and realizing that the dress she was wearing was not the one that she would have chosen to wear.
    There was a beautiful pink lace dress hanging in her closet that she wore for their fiftieth wedding anniversary that she loved so much. I had bought a blue one for myself and when I showed it to her, her eyes lit up and I knew that she wanted to wear one like that, without her even saying a word.
    I told her that there was a pink one at the store in her size and I would get the same dress for her and I would wear another style. She was so excited and looked so beautiful during that occasion. Both she and Dad were glowing with happiness after reaching that milestone in their marriage.
    My mother wanted all her personal greeting cards buried with her, which was the last thing that my siblings and I did for her. We tucked them around her, kissed her goodbye, and then the undertakers came into the church and closed her coffin.

I remember walking from the church to the cemetery and as I looked behind I noticed the large groups of people who were walking behind us. The road was filled from the church to the cemetery on this beautiful warm and sunny day.

I said to myself "this is a reflection of how my mother spread her love, kindness and infinite patience." All these people are here today to pay their last respects to a wonderful person who never had any room in her heart for hate or vindictiveness. Despite everything that she had to face in her life, she kept saying "there is good in the worst of us, bad in the best of us, so keep looking for the good in people and you will find it."

Back at the Waterford Hospital where the nurses recorded my mother's death, they gave two conflicting reports:

Nurse's Progress Notes
June 1st, 1995 0800-2000

Judi remains settled although still maintains that she had a reaction to Trinalin – refused to accept diagnosis of Bipolar illness.

Sister-in-law visited this morning with the news that Judi's mother passed away from a heart attack this morning.

Judi accepted the news fairly well. She cried for short period, but says she realizes her mother wasn't well and she's relieved her death was quick as her mother lived in fear of having to suffer another stroke.

Accepted this news and is coping well.

She was assessed by Dr. Karagianis and has gone on extended leave to attend mother's burial. She will return to hospital Monday. She promises to take her medications as prescribed.

Clinical Clerk's Note
June 1, 1995

Patient very upset today concerning mother's death. No somatic complaints.
Patient has been very stable this week. She is compliant with medications, not suicidal.
Impression: Stable
Plan: Accompanied extended weekend pass.

Nurse's Progress Notes
June 5, 1995 2300

Judi hasn't returned from leave.

Sister-in law in Carbonear contacted. She says that Judi will return in the morning. Her family are home from away and she will return with them tomorrow.

She states that Judi coped extremely well and read a passage at her mother's funeral. She states she has never seen Judi so calm and able to cope, although she is drowsy and tires easily.

Also feels that Judi is beginning to accept that there is something wrong as she overheard her tell a friend that she may have to take one of the medications for a long time.

Tuesday, June 6, 1995

I spent another few days visiting with my family after Mom's death and then I had to return to the cold, bleak, horrible Waterford Hospital. The days rolled into weeks of eating, sleeping, bathing and crying alone while taking drugs. I can't remember any person from the staff giving me one word of empathy, let alone sympathy for the sadness I was experiencing mourning my mother's death all alone and grieving like I had never done before.

I took accompanied passes as often as I could, but I was not free, as I was under the complete control of the Waterford Hospital and its employees. I had to conform to their forced drug therapy, even though I did not need them, as I had never suffered from the mood swings of bipolar disorder, and the drugs were making me worse.

I had complained of numbness in my head, neck, and fingers, constant fatigue, abdominal pain and feeling like I had "half-a-brain" as I could not concentrate or read. Finally I complained of blurry vision, before I started to become constantly nauseated and vomiting many times.

Nurse's Progress Notes
June 6, 1995 1600-2000

> **Judi returned from her pass in good spirits.**
> **She remained in her bedroom until her supper at 1730.**
> **Has been up and about unit since,**

2000-2400

> **Judi spent remainder of evening lying on bed.**
> **Feels drowsy and lethargic, but is much calmer.**
> **She feels she did very well at her mother's funeral and although she will miss her is able to cope with the loss.**
> **She would like to have the Haldol reduced because she feels she's unable to work feeling so drowsy.**
> **She is no longer saying she does not need the medication.**

Clinical Clerk's Notes. Rounds

> **Patient had a good weekend considering the circumstances, compliant with medications.**
> **She did mention fatigue with Haldol and asked for it to be decreased.**
> **As well, she feels she might get depressed with her mother passing away. In view of this, it is not advisable for her to leave the hospital at this point. She can have passes liberally however.**
> **She still lacks insight saying she will continue to take her medications, but does not believe she needs them.**

Plan:
1. **Decrease Haldol 3 milligrams, twice a day and at night, by mouth**
2. **Passes with family whenever necessary**
3. **Increase Lithium 900 milligrams.**

Clinical Clerk's Note

> **Patient seen briefly, agreeable to staying if she can have passes. She left for six hour pass.**

I was never agreeable to staying. I had no other choice: I was certified: Where would you think I would be better mourning my mother's death, confined to a room in a mental institution or with my family?

Nurse's Progress Notes
June 7, 1995 0800-2000

Judi was discussed in rounds. She would like discharge, but she was told that she should stay for a little longer.
Judi will admit that she's presently at risk for depression as she has found her mother's death very stressful. (Inaccurate)
Lithium increased to 900 milligrams at night.
Haldol decreased to 3 milligrams, twice a day and at night.
Judi has been complaining of drowsiness related to the Haldol.
She was on pass with her sister today.

2000-2400

Judi spent the entire evening in bedroom asleep. No interaction with others.

Clinical Clerk's Notes
June 8, 1995

Patient is distressed this morning and frustrated. She feels like she is being "destroyed" She used to be very energetic, always getting involved. Now she feels tired, has no interest in doing anything and is feeling depressed.
She attributes everything to Haldol, says her mother's death is not a stressor.
She believes if Haldol was decreased, she would be much better.
She is also complaining of numbness in fingers, neck and her head.
She is well dressed and groomed, behavior is appropriate. Tearful, crying during interview.
Mood subjectively and objectively depressed, affect appropriate.
No thought disorder, speech normal rate and volume.
No delusions/hallucinations. No suicidal thoughts. Still no insight.

Impression: dysthymic. (Any disorder of mood)
Plan: Consider decrease Haldol again; passes whenever necessary.

Nurse's Progress Notes
June 8, 1995 0800-2000

Judi very tearful at beginning of shift; upset because she was in hospital and taking medications which were causing her to "slow down" and states "only half my brain is working."
She inquired if writer could speak to her Doctor about discontinuing her Haldol.
During conversation, Judi continued to show no insight into her condition.
She stated she was here due to 'medication causing her psychosis prior to admission."
Patient presented many explanations for her many stressors, medication, ex-husband, extra course work at university.
Patient left unit at 1300 hours for a day pass with son to return at 1900 hours. Patient has not returned thus far.
Patient remains compliant with medications.

June 8, 1995 2000-2400

Judi returned from pass at 2030. She stated that she had a good afternoon. She requested that writer leave message that she would like a weekend pass to return on Sunday.

Judi also expressed concern that 'Haldol is not being decreased fast enough". She feels she is "doing fine on 3 milligrams and would like the same to be decreased to 1 milligram".

She continues to lack insight into illness.

She requests that writer speak to her doctor to tell him to decrease medication (Haldol) Informed Judi that her concerns would be noted, but she would have to speak with doctor concerning medications.

Settled in bed at 2200 hours.

Clinical Clerk's Notes
June 9, 1995

Spoke with patient this morning. She only took 2 milligrams of Haldol this morning and apparently stated she wanted it to be decreased to 1 milligram tomorrow, so she would be off it by Monday and would not have to come back to Hospital.

She keeps insisting she is well, and the Haldol is making her tired and unable to do things. She is asking for a weekend pass.

Patient was again explained her diagnosis and need for medications. As well she was told that her Haldol would be tapered off slowly, but it would be the doctor's decision, not hers. She seems agreeable after some persuasion.

She promised to be compliant with her medications over the weekend and seemed genuine.

She was well dressed and groomed and behaved appropriately. Her mood was subjective and objective euthymic, affect appropriate.

She was not irritable and in control.

Even though I had been provoked daily by unfair incidences, I have managed to keep calm and be patient.

Clinical Clerk's Note: cont'd
June 9, 1995

Her speech was normal in rate and volume, with no evidence of formal thought disorder.
No hallucinations/delusions. Good impulse control. <u>No insight</u>
Lab, Lithium 0.77.

Plan:

1. Accompanied weekend pass to ensure compliance

2. Consider decrease Haldol Monday, depending on her Lithium levels.

Nurse's Progress Notes
June 9, 1995 0800-2000

Patient seen this morning by unit team, concerning weekend leave/medication, and compliance.

Family friend accompanied patient on weekend.

The need for medication compliance was stressed to family friend and Judi. Patient to return to unit at 1000 hours Monday.

Clinical Clerk's Notes
June 12, 1995

Patient returned from weekend pass this morning. She had a good time with her family, but complained of being tired and having no energy. She said she would sleep a lot during the day. She also complained of abdominal pain in the evenings.

She is asking for Haldol to be decreased and hoping for discharge soon.

She is compliant with medications even though she still believes there is no need for them.

She is well dressed and groomed, does appear fatigued. Behavior is appropriate, good impulse control.

Mood is subjectively and objectively euthymic, affect normal, speech normal rate and rhythm. No evidence of formal thought disorder.

Orientated times three. No suicidal ideation. Insight remains nil.

<u>Impression:</u> Stable on medications.

<u>Plan:</u> Consider decreasing Haldol to 2 milligrams twice a day.

<u>Resident's note</u>

Agree with above note. Will decrease Haldol to 2.5 milligrams, twice a day.

Nurse's Progress Notes
June 12, 1995 0800-2000

As stated above. Judi has spent most of the day lying on her bed. She seems lethargic and slow. Feels sleepy all the time and says the slightest amount of activity tires her out.

She is unable to take daily walks like before she started medication.

She is asking about being able to drive her car and would like to speak with doctor about this in the morning.

She is gone for walk with female co-patient this evening.

2000-2400

She had refused 1800 medication with request to have them at 2200 hours to postpone sedative effect.

She did comply with medication at 2200 hours.

She was only up on unit for short period this evening, in her room for most of the time.

She continues to complain of feeling sedated.

Clinical Clerk's Notes
June 13, 1995

Patient doing fairly well. Still complaining of fatigue. She takes her medication in the evening to minimize the sedating effects.

She is concerned about long hospital stay and hoping for early discharge. Concerns re future.

Well dressed and groomed, behavior appropriate. Good eye contact. Spontaneous smile, spontaneous speech. Speech rate and volume normal, no evidence of formal thought disorder. Mood subjectively and objectively euthymic, affect stable. No hallucinations/delusions, no suicidal ideation.

Laboratory. Lithium 0.88

<u>Impression:</u> bipolar disorder, stable

<u>Plan:</u>   1.   unaccompanied passes
2.   Early discharge this week

What I find very interesting is that I was allowed to drive my car. I had asked if I were allowed to drive it, and to my surprise, the authorities said "Yes." I needed to get my car inspected, so I made an appointment at Dave Tucker's Subaru and drove to Topsail Road and chatted with Dave while my car was getting the oil changed and tune-up.

I also drove to my cousins, Gary and Barb Davis' house and had a visit with them on my way back to the Waterford Hospital. Can you believe this occurred when I was "certified" into a mental institution, against my will? Does this make sense?

I often wonder if the police had stopped me and I told them I was a certified patient at the Waterford Hospital, what would have happened then? What would be the result, if I were unfortunate enough to get into an accident?

It probably was a bad idea to drive, but I felt I was competent enough to handle the car at the time, but I was very fatigued and had little ability to concentrate. I drove slowly and carefully, but I was definitely incapacitated on drugs. Perhaps also I just wanted to have a little control over my life again, but it should not have been behind the wheel of a car.

Nurse's Progress Notes
June 13, 1995 0800-2000

Judi spent a quiet day up and about unit. No complaints voiced.

2000-2400

Up and around unit until 2100, then settled to bed. She continues to lack insight and insists her admission was precipitated by a reaction to antihistamine.

She is complaining of the inability to read and hopes she will be able to stop taking Haldol soon or at least have the dosage reduced again.

Judi complains of finding the time long here and spoke of requesting an extended leave when she next speaks to doctor.

She continues to complain of sedation. Spontaneous, euthymic mood with appropriate activity level.

The physician cannot prescribe by letter, he must feel the pulse. / Seneca

My mother, three days before she died. The picture was taken by my brother Jim. I kept it in my view, during my fifty five days in the courtroom.

Chapter 10

Discharge from Waterford Psychiatric Hospital
June 19, 1995

Then one day, June 14th, Dr. Karagianis told me I was going to be discharged after an extended weekend pass. (The morning I was supposed to go home, I became violently ill with nausea and vomiting and I was told I had to stay in hospital while I was vomiting.)

I was sick in bed for three more days, completely nauseated and vomiting up anything that I drank. I had absolutely no nursing care, no personal hygiene, no comfort, and no nourishment, except there was a large bottle of ginger ale and a glass placed by my bed side. I managed to get to the bathroom to relieve myself and otherwise nothing else was done for me and I could do nothing for myself as I was too ill.

I was told I had to not vomit for twenty-four hours before I could be discharged, so I kept my vomiting quiet and did not tell anybody. I actually hid in the bathroom when I was nauseated and lied to the nurses that I had not vomited.

Clinical Clerk's Notes Rounds
June 14, 1995

Patient has been stable this past week and her Haldol has been gradually decreased. She is tolerating this well with no relapse of her manic symptoms.

There are no major side effects aside from abdominal pains that settle when Lithium is taken with food.

She is still complaining of drowsiness, which has been improving with decreased Haldol. She is experiencing no EPS. (Extrapyramidal symptoms are abnormal twitching or other uncontrolled movements that sometimes happen to patients who consume these neurotoxic drugs. These drug induced motor abnormalities may become permanent.)

Patient seems to be doing quite well, hoping for discharge.

For now, the Haldol will be decreased and patient will be reassessed later this week.

Impression: bipolar Illness. Stable

Plan: Decrease Haldol I milligram by mouth, twice a day, and at night

Clinical Clerk's Notes
June 14, 1995

Patient seen briefly this morning with Dr. Karagianis.

She is still feeling lethargic and complaining of being unable to concentrate, and blurred vision. She was told this would improve as medication is tolerated.

Agreeable to long weekend pass, reassessment on Monday and possible discharge then.

Mental Status Examination: Grossly normal
No mood symptoms.

Plan: Extended weekend pass.

Nurse's Progress Notes:
June 14, 1995 0800-2000

Judi spent the morning up and about unit. Left on pass at 1230, supposed to return at 1630 but returned at 1700 hours.

She took her 1800 hour medications

2000-0800 Spent quiet evening on unit. No complaints voiced. Meals and medications taken well. Settled at bedtime.

Clinical Clerk's Notes
June 15, 1995

Patient feeling nauseous today, vomited times two since this morning. Not complaining of abdominal pain, chest pain, or shortness of breath.

She is thirsty, but voids five to six times a day only.

Patient is considering staying in this evening and leaving tomorrow morning for weekend pass.

On exam, patient looks unwell, fatigues and slightly pale. Afebrile (without fever). No tremor. No other signs of Lithium toxicity.

Mental Status Examination:

Little disheveled, behavior appropriate, mood subjectively and objectively euthymic, affect appropriate. Speech normal rate and volume. No evidence of thought disorder. Not suicidal.

Showing insight concerning present problem and agrees to stay until well enough.

Impression:
1. Viral gastroenteritis
2. bipolar Illness stable

Plan:
1. Leave for weekend pass tomorrow
2. Repeat Lithium levels in morning.

Nurse's Progress Notes
June 15, 1995 0800-2000

Judi complained of not feeling well this morning and has not eaten today. Vomiting times one. She was seen by Dr. Kennedy and repeat Lithium level ordered for tomorrow morning.

Patient decided to stay and see how she does overnight- look at weekend leave tomorrow

2000-2400

Judi spent the evening in her bedroom. Says she is feeling a bit better. Vital signs were done and charted. She did not eat any bedtime snack.

Clinical clerk's Notes
June 16, 1995

Patient seen this morning, She is still not feeling well. Vomited during the night. She cannot keep fluids down. She agrees to stay and if she feels better and has a twenty four hour period with no vomiting, she can go on her weekend pass.
Lithium not given last evening in light of her present condition
Today she is still showing no other symptoms of Lithium toxicity. Lithium level 0.52.

Impression: viral Illness.

Plan:
1. Can have weekend pass if feeling better
2. Decrease Lithium while vomiting to 600 milligrams.

Nurse's Progress Notes
June 16, 1995 0800-2000

Judi has been feeling ill today. She feels lethargic and is unable to keep fluids or solid foods down.
Ginger ale and juices supplied to her.
Gravol 30 milligrams intramuscularly given at 1630 hours.
She had a bath and bed linen changed.
Lithium level 0.52 millimoles/litre today. Lithium decreased to 600 milligrams at bedtime.
She became agitated at 1600 hours. States she felt like she is going to die. She requested transfer to a general hospital. She states she feels she needs intravenous therapy. Her skin turgor is good. Seen by Dr. Pratt.

I know that I became agitated "after" Gravol was injected in me. Why Gravol would be ordered and given to me is a question that still needs to be answered. "Allergic to Gravol" was written on my medical records, and I had an allergy bracelet on my arm. Nevertheless, I was ordered Gravol and injected with it, despite the safety nets that were supposed to be in place.

I had not taken any Gravol since 1983, when I had had an adverse reaction to it. It had made me hyper and agitated. At that time, I should have been more careful when I took Gravol with other antihistamines as there is a warning against that practice in the Compendium of Pharmaceuticals and Specialties (CPS).[1] According to CPS, there are reports that Gravol can cause hallucinations after taking doses of 500 and 700 milligrams. Being so sensitive to drugs, I would have had an adverse reaction to a much lower dose. Also I was prescribed Halcion by Dr. Jain on top of the Gravol and antihistamines, which I unknowingly took.

Nurse's Progress Notes Cont'd
June 16, 1995 2000-2400

Judi spent most of the evening in her room lying on her bed. No complaints voiced.

---

[1] Compendium of Pharmaceuticals and Specialties (CPS), 2008 pg. 1023

Nurse's Progress Notes:
June 17, 1995 0800-1600

Up early this morning. No complaints of nausea or vomiting this morning. Judi says she has not vomited in twenty four hours.

She left at 0900 hours for weekend leave and took medications with her, to return Monday morning for Lithium level.

Clinical Clerk's Notes
June 19, 1995

Patient came back this morning. Spent a good weekend. She is still feeling nauseous, but has not vomited since leaving the hospital.

She noted change in bowels, there was mucous, but no blood, not loose

Appetite, sleep intact, energy still down. She was compliant with medications.

She was well dressed and groomed, behavior appropriate. Good eye contact. Speech volume and rate normal. No evidence of formal thought disorder.

Mood objectively and subjectively euthymic. No delusions or hallucinations. Not suicidal. Patient will come back in morning to see Dr. Karagianis.

Impression: Stable
Plan: Discharge today.

Dr. Karagianis wrote:

Still no insight, but no active signs of bipolar disorder. Pass went well. Discharge.

Nurse's Progress Note
June 19, 1995 1530 hours

Discharged today to home. Follow-up appointment for Terrace clinic given.
Condition improved as noted above. Prescription given.

Finally, on June 19th, they let me go home. By that time, I was complaining of blurred vision, which was the last complaint recorded on my medical record before I started vomiting. This was a serious side effect of lithium neurotoxicity that should have been recognized and comprehended that something was seriously wrong. Now I was more neuro-toxic than ever.

I did not know what caused me to become so ill until my charts were studied by a psychiatric specialist Dr. Eleanor Stein, who correlated my lithium blood levels to the times I began vomiting. Each time the levels went to 0.7mmol/l, I was lithium toxic, which would induce vomiting. The doctors ignored my symptoms, and wrote on my records that they felt the vomiting was psychogenic, or brought on by me, if not a viral illness. According to their ignorance, I was not lithium toxic, as my blood levels were within normal range.

It appears that the physicians caring for me had no concept of individual drug sensitivities and did not realize, due to my undiagnosed illness of cfs/fm/mcs, that I was indeed neuro-toxic at a lower-than-normal blood level.

It is also well known that people with fm/cfs/mcs can only tolerate one quarter to one tenth what a normal person can tolerate when prescribed drugs.

When I reviewed my records, I also noticed that the responsibility of my complete medical care was facilitated, orchestrated and recorded by a clinical clerk on her first rotation to psychiatry, with very little supervision by my treating physician, Dr. Karagianis or his resident. It is on record that Dr. Karagianis had only visited me four times during the five weeks of confinement.

I have counted sixty-three times on the nursing progress notes that I had complained of side effects of the drugs that were prescribed, and nobody acted to evaluate the situation further for an error in diagnosis and treatment. They had treated a diagnosis of bipolar disorder, and had ignored my actual complaints of the symptoms, which I was voicing and displaying to them.

A specialist is a doctor who trains his patients to become ill only during office hours. /Anonymous

Bowring Park

Chapter 11

The Summer of 1995

After I went home, I kept taking the lithium at a reduced dosage since I could not tolerate what had been prescribed as I would become nauseated and vomit it up anyway. Even taking reduced amounts, I did not have the strength or energy to do anything. The least bit of activity tired me out and I slept sixteen hours a day. My life was reduced to a vegetative state, even though I wanted to do the things I did before, I could not.

A few days after I was discharged, I attempted to go back to work and I was accepted back as if nothing at all had happened to me during the previous six weeks, and I did not feel any discrimination or treated any differently by anybody! The big problem I had though was that I had absolutely no short term memory left. My concentration was minimal, and I was always fatigued.

I had the constant feeling of being nauseated and felt as if I had a perpetual flu-like illness. All I wanted to do was lie down.

I had to write notes to myself to remind me that I had written a note about matters that were important for me to do. I became confused quite easily, and due to low levels of concentration, I could not retain anything that I would read. I could understand what I was reading, but could not retain it.

I was very thankful that my writing skills were not affected. If anything, they were improving as I kept writing in order that I could refer to my notes to remember what was happening to me as I was having more and more difficulty with my short term memory.

I was so mentally and physically fatigued that I could not function to the capacity required to perform my job effectively and it made me very frustrated and anxious.

My life consisted of going to work, coming home, eating and going to bed, so that I could have enough energy to get up and go back to work the next day. After struggling like this for a week or two, I had no other choice but to take sick leave again, knowing I would have no income. There was no way that I could work with my compromised mental faculties.

Within two months there had been quite a change in my intellectual ability, going from university material at a Masters level to a moron. Any fool should have known that it had to be drug induced from taking Haldol and Cogentin, plus Lithium.

I spent the rest of the summer just existing. I moved out of the apartment in my son's home and rented a room in my friend Tony's house, with two other boarders. I did not feel depressed, but very sick and tired. The smallest activity would exhaust me!

There was a single tennis court in the basement, as the children were avid tennis players and had won many competitions. I loved playing the game; but I could not last any more than a few minutes before I would become exhausted and have to stop playing. Everything I did exhausted me! I was in excellent physical shape while going to university in the spring of 1995. Now just three months later I was an invalid. I blame my state of health on the incorrect drugs I had been prescribed. This had made my condition of undiagnosed fm/cfs/mcs worse.

The vomiting that I experienced, especially when I rode in a car, had been recorded as psychogenic, and the constant nausea, lethargy, memory loss, dizziness, ringing in my ears, blurred vision, loss of balance, and sleeping sixteen hours a day, was attributed to depression. I knew I did not feel the gloom, doom and the darkness of depression that lead to thoughts of suicide. I felt drugged to incapacity, and simply wanted somebody to revisit my diagnosis and treatment.

Every day, I was struggling very hard to live! I never once thought of suicide, or wanting to end my life. I remember the staff health physician, Dr. Jean Griffiths, asking me that question when I was going through my second marriage break-up. I said "Dr. Griffiths, I want to live, not die, but I want to feel better than I do now."

I kept my appointments with Dr. Karagianis, and around the end of August, he added Prozac 20mg to take with the Lithium. I had never taken such a high dose of Prozac, so I took it reluctantly, as it made me restless, anxious and jittery.

There were times that I would just have to leave the presence of other people and keep to myself as I felt so strange, different and disconnected. This was not at all like I was normally, which was usually very sociable as I have always liked being around people. I felt the Prozac was causing these reactions, so I did not take it every day.

I brought my friend Eleanor with me when I went to see Dr. Karagianis since she knew him personally from going to the same church. I thought that she might be able to convince him that something is seriously wrong here, but that backfired.

Somehow he must have gotten the impression from Eleanor that I was a little better lately than I had been when I was first discharged from the Waterford Hospital and before I started taking Prozac. Therefore he ignored everything I was telling him about the way I was feeling. These symptoms were serious side effects of the drugs I was taking for the treatment of bipolar disorder. By this time, I was really caught in the psychiatric trap, and too physically ill to even know I was there. I had turned into a zombie!

At this point, I did not know that I had had a problem with my thyroid gland even before Lithium was started. During the Waterford admission, my thyroid hormones were checked through blood work. The results came back ten days after I was admitted showing that the Thyroid Stimulating Hormone (TSH) was 8.2. Normally this reading should be 3.5 to 5, so mine was well above normal, indicating my thyroid gland was sluggish. Lithium is well known to negatively affect the thyroid gland, therefore is there any wonder I was very fatigued and sleeping over sixteen hours a day while taking Lithium? Now Prozac was added to the mess!

That summer I met a pharmacist and we befriended each other. My first husband was a pharmacist and my friend knew him professionally. I was too exhausted to have very much energy to put into a relationship, however, we spent some quality time together and I really appreciated his friendship. I could talk to him about my illness and the drugs I was prescribed.

We went for drives together and I would be so nauseated and tired that I must not have been the best company. I never lost my sense of humor though, and he commented so many times about how I always made him laugh. I could see he was getting serious with me, but at that point in my life, I had nothing to offer anybody. I was using every ounce of energy trying so hard to just survive each day and be as happy as I could while feeling so ill.

I severed our friendship and we went on with our different lives apart, not even being friends. I guess it was bad timing that's all, but among my treasures, I have kept the beautiful poems that he gave me that summer of ninety five. Here is the poem I wrote him.

Good Friends

It was I who chose to be with you,
many months ago.
When I asked to be your dancing mate,
our faces all aglow

I did not want to fall in love
I made that very clear.
I need some time to rid myself
of pain that's hard to bear.

The signs were showing in your face
of love I can't return.
It's wonderful to feel that way,
together, but not alone.

You are man of character,
Intelligent, and kind.
A girl can be so honored,
to absorb your heart and mind.

I didn't mean to make you sad,
or hurt in any way.
I hope we'll always be good friends
And trip around the bay.

We will always dance together,
The very best of friends.
I really want to be alone,
until my body mends.

If not, I will be leaning
on someone else for strength.
I'll never try to stand alone
or straighten up what's bent.

I was still consulting with Dr. Karagianis on a monthly basis. These could not be described as 'visits' as they were so impersonal and lacked any human connection. Here is his interpretation of my symptoms of profound fatigue and feelings of being unwell.

Dr. James Karagianis
Letter to Blue Cross
November 15, 1995

Re Judith Day
DOB, October 27, 1945

I am writing in response to your letter requesting further information on the above named patient.

I first met her on May 15, 1995 when she was admitted to the Waterford Hospital. Her family reported that in a week prior to admission she developed an elevated and irritable mood and showed increased energy.

(This was untrue. I was physically ill with sinusitis and bronchitis.)

She had increased self-esteem and was quite grandiose and excessively talkative. She also noted poor sleep at that time and was doing impulsive things. The day prior to admission she thought she received a sign from God. (I was not sleeping due to blocked sinuses.)

At that time she had no insight into her illness and therefore she had no complaints. She was treated successfully with Haloperidol 1mg. twice a day and Lithium 900 mgs at night and was discharged June 19, 1995 with a prescription for these medications. (I had insight that I was having a psychotic drug reaction.)

Of note while she was in the hospital she displayed grandiose delusions, agitation and aggression, and she required certification. The final diagnosis for this admission was bipolar disorder, manic phase.

(I was certified upon admission and was shocked when I was informed that and had to take medications or be injected.)

Psychiatric history reveals that her first admission for psychiatric reasons was at the Health Science Centre about ten years ago. At that time she had religious delusions and had threatened family members. The old charts were impossible to read so the diagnosis is uncertain for that admission, but the patient insisted that that episode was induced by Gravol. (and antihistamines plus drugs (Halcion) Dr. Jain had prescribed for me.)

It is my impression that was the first episode of bipolar disorder but she, herself, believes that this was not the case. Between that admission and this most recent one she functioned relatively well as was able to perform her job as a nurse. (True.)

I have not performed any psychometric tests since none were indicated.

Past medical history is non-contributory. However she believes that she is sensitive to medications and she attributes her recent relapse to a reaction to the medication, Trinalin. (There was no medical history taken or recorded on my psychiatric records at Waterford Hospital and the elevated TSH was ignored.)

Since discharge I have seen her five times. She switched into a depressive phase which began soon after discharge. Symptoms at that time were poor appetite, low energy, feelings of hopelessness and depression, with objectively depressed affect and tearful expression. She was quite lethargic at that time, which was July 4, 1995.

(I had never felt hopelessness and depression.)

When I saw her again on August 24, 1995 she was still depressed despite reducing the Lithium and Haldol the previous visit. At that point I discontinued her Haldol and added Prozac 20 mgs. Once a day and continued the Lithium. She indicated she was compliant with this, although her levels were a bit low. (I was not depressed.)

When I saw her on September 6, 1995, she was feeling a little bit better and was able to get out of bed a little easier and was able to go for walks and felt less hopeless and less gloomy. Her crying spells were continuing. Her sleep was a little better. Objectively at that time her affect was less depressed but still not back to normal. She was not tearful at the time of that examination. (I was profoundly fatigued.)

I saw her again on September 27, 1995, and she said she was "struggling to live." She was staying in bed until noon and taking a very long time to get going during the day. She still felt depressed and had to push herself to do anything. Her appetite was still below normal, but she would crave sweets. She complained of poor energy and excessive sleep, i.e. fourteen hours a day. She reported fewer crying spells at that time.

Objectively during that assessment she appeared a little brighter and was not tearful but was able to smile. I felt she was slightly improved and I reduced her Lithium.

The next and last time I saw her was October 10, 1995. At that time she says she was able to do a little bit of entertaining and was able to go bowling; however she was quite sore afterwards. She felt she was still struggling but less than two weeks ago. Her sleep was a little more normal. She didn't feel depressed at that time and did feel more hopeful. However her tolerance for activity was still below normal.

Objectively her affect was a little brighter and she did not appear depressed. There were no tears. She appeared optimistic for the future. I requested that she should make an appointment for two or three weeks from October 10th, but so far she has not turned up and I believe she had an appointment which she missed.

Her current treatment includes Lithium 300mgs. HS and Prozac 20 mgs. once a day.

The last time we met I did not ask her about whether she thought she had bipolar disorder, so I cannot comment on her current insight. She was alert, orientated and in no distress. Her clothing was casual. Her hygiene was good. Her behavior was appropriate. Her mood was "not depressed." Her affect was euthymic. Her thoughts were normal in form and content and there was no evidence of any hallucinations, no delusions, and no suicidal ideas. I did not assess her memory.

I believe her prognosis is good for returning to work. Unfortunately, I did not find enclosed in your letter a description of her job as a registered nurse.

When patients with psychiatric conditions are forced to go back to work, before they are ready, this is a considerable source of stress and may contribute to a relapse and even re-admission and further time loss from work. Work is very important to Judi and I am confident that she will return to work, when she is ready.

She had already tried to return to work unsuccessfully at least once during this episode, and I think, it will set her back further. In summary, I believe she is recovering from a depressed phase of bipolar disorder. If she takes the recommended treatment the prognosis is good.

I hope this report is some assistance to you.

Sincerely yours,
James Karagianis, MD FRCPC Psychiatrist

I was extremely tired the day in November, 1995 when my father phoned me for a copy of the poem "Faith" that my late mother had written for me when I was in the Waterford Hospital. He needed it as soon as possible as he wanted to include it in a booklet of her poetry that he was getting compiled. I had the only copy of that particular poem so I pushed myself very hard that day to find the poem for him as I wanted it included with the rest of her poetry.

I had exhausted myself completely, and the next day I was very confused and disoriented. I was so tired that I was wired. I had gone into the wrong bedroom during the night, woke up Tony and possibly scared him. I remember sitting at the kitchen table with the other boarders and talking craziness. I was not feeling well at all: I was shivering cold, confused, restless and hyper.

I phoned my friend Eleanor to ask her to take me to the emergency department of the Health Sciences Centre. I knew I needed to be supervised again before the possibility of blacking out and going psychotic might occur. I was admitted there in the psychiatric ward for two weeks under observation and I was now getting deeper and deeper into the psychiatric trap.

He who conceals his ills cannot expect to be cured./ Proverb

Emergency Entrance Health Science Centre

Chapter 12

Admission to Health Science Centre
Nov. 27, 1995 - Dec. 12, 1995

During the admission to the Health Science Centre, it was recorded on the physicians' assessment:
   "Patient quite agitated, rapid speech, tremors. No DLROW, which meant I could not spell "world" backwords.
   <u>Memory</u>: remembered one of three items at three minutes.

Nursing Assessment
November 27, 1995 1330 hours

Appearance:
Neatly dressed, well groomed. She appears stated age. She was lying in bed and appeared relaxed- patient had received Ativan 2 milligrams around one hour prior to interview due to increased pacing and increased thoughts.
   Tired looking

Behavior/speech:
Spoke coherently in a normal tone of voice. She maintained good eye contact. Prior to receiving Ativan, her speech was pressured.

Affect:
Appropriate to mood.

Perception:
Denies any auditory and visual hallucinations.

Cognition:
Memory, both short and long term memory appear intact.

Orientation:
Orientated to person, place and time.

Judgment:
Would mail the addressed, stamped envelope if she found it.

Insight:
Appears fairly good.

Abstract Reasoning:
Able to abstract proverbs

Thinking:

Thought content: No evidence of any formal thought disorder during interview. She states her thoughts are sometimes "speeded up". Stated to other patients she was God, prior to receiving Ativan. She stated "I know I'm not God."

Lethality
Self: no others: no

Privileges on admission

Closed observation
Patient's goals and expectations of hospitalization:
Not to go psychotic. To improve each day. She is hopeful she will only be here for a short period, two or three days.

Patient states she was increasingly "hyper" and "came into hospital before I reach the point of outburst."

The impression again by the physician was that I was manic. It was not noticed that as soon as the drugs were discontinued, I returned back to a normal mental state.

The Prozac was discontinued and then Haldol and Cogentin were added again to the Lithium, which was increased up from 300 milligrams daily to 1200 milligrams, and caused considerable side effects.

Excerpts from the journal:

Monday, November 27. 1995

I am admitted to HSC Psychiatric ward. I have been feeling different the past two days, so I thought it was better for me to come to hospital and be treated before I became too sick.

The nurses and the doctors have been super to me. I feel so safe here and I know with the proper drugs I will be okay.

Boy! Do I know differently now, these drugs were killing me.

(I have been walking the corridors trying to burn excess energy and I feel so good about the fact that before long I will be home again all well. I ate supper, terrible.

I talked with Dr. Allan Kwan as he strolled by the dining room on the way to the desk. He was very nice to me.) Roxanne Thornhill came and took my blood. She is Kim's, (a coworker), sister-in-law. B.H. came and introduced herself to me. She has been a patient here for over two months.

I went down to the lounge and had a chat with some of the patients. I met C., Dr. Karagianis' patient from Gander. She has a terrible guilt over her mother's death some twenty three years ago. She has been on sleeping pills ever since. She is trying to withdraw from them. She appears very tormented. I told her to speak to Dr. Karagianis about the feelings of guilt, but she says he never has time to listen to her.

I laid on the bed and slept for a little while when Kathy, another coworker, came in with her sweater she is knitting.

Nurse's progress Notes
November 27, 1995 2140 hours

Maintained on close observation. During the early evening Judi was talking with other patients in the lounge. She is noted to be exhibiting pressured speech and started to pace more and more. She approached other patients and stated she was "God". She was encouraged to rest and was given Ativan 2milligrams sublingually at 1815 hours.

She settled well, felt relaxed. No pressure of speech noted.
She stated "I know I am not God" Cooperative and pleasant. Tired looking
She has been sleeping since eight thirty.

Tuesday, November 28, 1995

I had a good night's sleep after taking a sedative. I awoke around eight and had breakfast. I was walking around the corridors and sat by the desk. While there, I overheard the nurse discuss about changing my doctor from Dr. O'Loughlin to Dr. Craig. That made me very upset because they did that without asking me. I went to the desk and told the nurses that I do not want Dr. Craig.

Dr. Craig just interviewed me and I told him that I do not want him as my doctor, as I need a female psychiatrist such as Dr. O'Loughlin. He told me I had no other choice since I was assigned to him. He told me I would have to discharge myself, and be admitted under Dr. O'Loughlin when she is on duty. I told him I would do that, but I couldn't.

I did not discharge myself, as I felt I was too ill, and needed somebody to listen to me, as I felt the Prozac had caused this new episode. I wanted to discuss this with someone who I thought would listen. I knew Dr. Craig from working at the Health Science Centre, and I did not particularly care for his attitude and heavy-handed manner that he displayed with his patients.

Dad came in to see me. We had lunch together and I reassured him that everything is okay and I should be discharged in a few days. I played pool for a short time and came back to my room for a rest.

Wendy, the nurse who is relieving me at work, came to visit me. I appreciated that very much. I showed her some of my papers on health care issues, and told her what I had planned to do with them.

After taking a bath, I went down to the TV room to watch a movie. I was so sleepy I had to leave and go to bed. I slept for two hours and did not wake until nine in the evening.

After I slept, I walked down to the lounge and read the newspaper. Then K.B. played a game of ping pong with me. After about ten minutes of playing, I was exhausted with perspiration rolling down my forehead. It was nearly bedtime so I decided that I had better slow down and relax so I could sleep tonight.

I went to bed around eleven and I slept well except I was awakened many times when the nurses were making their rounds. I went back to sleep though, and slept until seven thirty. I will get dressed now and face today.

Nurse's Progress Notes
Tuesday, November 28, 1995

Patient remains on close observation.
She was awake at change of shift last evening, but asleep by 0030 hours.
Patient appears to have slept well overnight. Currently still asleep.

Clinical Clerk's Notes
Tuesday, November 28, 1995

Case presented. Known longstanding history of bipolar disease.
Plan, Discontinue Ativan and Prozac, continue Lithium and Haldol.
Met Ms. Day with Dr. Craig and myself this morning. Patient refused to talk with Dr. Craig. Patient wants to talk with Dr. O'Loughlin.
Option given --- To see Dr. Craig in hospital, or self discharge.

Dr. Craig's Clinic Note:

Ms Day refused to speak with me, thereby restricting the observations I was able to make personally.

Nursing Progress Notes
November 28, 1995 at 1545

Judi has been maintained on close observation during the day.

She was transferred to S/O Dr. Craig today and had expressed much displeasure re same because she prefers to be treated by a female physician.

Throughout the morning she was noted to be, question, counseling another female patient as well as she made phone calls to the Housekeeping Department to clean the windows in the lounge.

I did not call the Housekeeping Department. What actually happened was that John, who worked in the operating room for the department of 'Housekeeping" came to visit me and I mentioned to him that the windows in the lounge were so dirty that you could hardly see through them. We thought it would be nice for depressed people to be able to have a better view of the outside and perhaps it would encourage and motivate them to feel better. John, who respected my opinions and ideas, observed the state of the windows and arranged for them to be cleaned.

Nurse's Progress Notes cont'd
November 28, 1995

She napped after lunch for a short while.

During the interview this afternoon she denied having weird thoughts and mood noted to be quite labile (at times tearful).

She is noted to be quite guarded in her conversation with some evidence of question; delusional thinking- continues to talk about messages on answering machine.

She also expresses some concern about the possibility of being "certified", but did not want to elaborate.

Nurse's Progress Notes
November 28, 1995 2200 hours

Maintained on close observation
Received several visitors during early evening, good interaction noted.
Spent some time watching T.V., She napped for a short period.
Overall appears more subdued this evening.
Tired looking. States her thoughts are clear and admits to feeling tired.

Every little thing I did or said is recorded in a way to reek of abnormality

Wednesday, November 29, 1995

I have a terrible looking sore on my middle finger right hand. I cleansed it with Hydrogen Peroxide and put a band aid on it. I hope that it will begin to heal now.

I went down to the lounge for breakfast, but did not eat much. I played solitaire and became very sleepy, so I came back to my room for a nap. When I came back to my room, I began reflecting and comparing the two psychiatric admissions.

Waterford Hospital	*Health Science Centre*
Most of the nurses had no time for the patients and many times I saw neglect.	*The nurses are very caring towards the patients, and are there to meet their needs.*
Most of the nurses never smile at the patients or listen to their needs.	*The nurses smile at the patients and are very courteous to them.*
Patients are neglected.	*Patients are made to feel at home.*
Nurses were very scarce there most of the time.	*There appears to be more nursing staff here.*
It took two people to dispense the drugs most of the time.	*It only takes one nurse to do that task here.*
The nurses are not in the kitchen when meals are served, elderly people have to pour hot water from very large kettles into Styrofoam cups which is very Unsafe.	*There was a nurse in the area while people were eating.*

Some of the patients went for a walk with a nurse. I was not allowed to go as I am under supervision. Hopefully tomorrow I will be able to go. I am really sleepy now since I took my medication. I played several games of darts, but it tired me out completely.

My friend Bev visited until it was time for dinner. After dinner, I took a hot bath and washed my hair. Now I am waiting for the news to come on TV. I listened to the news, and, then I watched only part of a movie, because Eleanor and Blanche came to visit me, so I came back to my room to talk to them.

It is now nine thirty, and I will soon be getting ready for bed. I know I will sleep well tonight because I feel so tired.

Nurse's Progress Notes
November 29, 1995 0616

    Continues on close observation.
    Slept well-awake briefly at 0500 hours

1420

    Stated thought processes- Described having had racing or "weird" thoughts on day of admission. However feels much better since admission. Reports decreased racing thoughts. No pressure of speech.
    However was very controlled during interview. Little range of motion.
    Little insight into illness, doesn't believe she has bipolar Disorder.
    She is attending all of unit's activities. Continues on close observation.

Clinical clerk's Note
November 29, 1995 1530

    Patient states she has no "strange' thoughts or feelings today.
    Feels much better, mood good, appetite good
    Energy decreased, sleep increased. Feels she is sleeping too much.
    She is asking for medications to be changed re decrease Haldol, discontinue, it makes her sleepy and feels Lithium has stopped her thoughts from racing.
    Patient still has pressured speech and theatrical in postures.

She denies hallucinations/ delusions.
She denies feeling high or depressed.
Patient is more accepting to have Dr. Craig as psychiatrist.

<u>Plan:</u> Discuss with Dr. Craig

1. **Medications, increase Lithium tomorrow, question.**

2. **Patient requesting pass for weekend.**

I had no other choice but to accept Dr. Craig as my treating physician. I couldn't self discharge as I knew I needed medical attention, but was getting further into the psychiatric trap.

I needed a psychiatric specialist to assess my medical history and try to figure out what is going on with me because I was now on medical sick leave due to a psychiatric diagnosis and would never be able to return to my job without having the proper diagnosis and treatment that I could tolerate.

Lithium was making me nauseated and Prozac was making me crazy. Both together were making me worse, as I was becoming more weakened and fatigued daily. As long as I was on sick leave from my job at Health Science Centre, I had to conform to the medical treatment that the doctors prescribed, otherwise I would not qualify for any sick benefits.

I had no other choice but to take as much Lithium as I could tolerate as the blood Lithium levels were checked regularly and that level was all that mattered to those psychiatrists. If my thyroid hormones were checked as well, they could have possibly prevented my compromised physical state from deteriorating further.

Despite a normal mental status being recorded on my chart, the plan was to **increase** Lithium tomorrow.

Nurse's Progress Notes:
November 29, 1995 2120

Maintained on close observation. Received female visitor earlier in evening and male visitor later – Good interaction noted. Had tub bath. Patient watched T.V. off/on throughout evening. Overall mood is quiet.

Thursday, November 30, 1995

I woke up early in the morning and I decided to write down some of my thoughts. I believe I could go home now as my mind is not racing like it was three days ago. I believe that I will be alright now.

Today is Winston Churchill's birthday. He was also a man that the psychiatrists in retrospect have diagnosed as bipolar disorder because of his dark feelings during the Second World War. With the responsibility of being Prime Minister of England, do you make any wonder that he had his ups and downs? Would he have functioned as well if he were forced to take Lithium and Haloperidol? I don't think so.

His true importance, however, rests on the fact that by sheer stubborn courage he led the British people and with them, the democratic western world from the brink of defeat to a final victory in the greatest conflict the world has ever seen (B.L.B).

Florence Nightingale's malaise was considered psychiatric as well for a time, but now her symptoms fit more with fibromyalgia and chronic fatigue syndrome, and her birthday, May 12, is the International Fibromyalgia Day.

I wonder how many bright and creative brains are going to be ruined in the future by psychotropic drugs. I know they have messed up mine.

I got up and had my blood taken at seven. I then took my bath and got dressed. I feel good but the least little bit of exertion drains my energy. By the time I had my bath and did my hair, I was exhausted. I am ready to go back to bed again.

There was a group meeting at 1000 that I attended. After the meeting, I spoke to Dr. Craig for a second and then we went for a twenty minute walk. Brian visited me and chatted to Dr. Craig.

It is now 2200 and I am getting ready for bed. I feel I will sleep well because I did not nap at all since I woke at 0700.

Nurse's Progress Notes
November 30, 1995 0600

Close observation maintained. Asleep at change of shift appears to have slept well.

1500

Has attended and participated in all unit activities today.
She continues to describe her mood and thoughts as good.
Would like her Haldol decreased and she has discussed same with Dr. Craig.
No pressure of speech or signs of agitation during conversation. She remains limited in her understanding of Bipolar illness.
Continues on close observation

Clinical Clerk's Note
November 30, 1995

Patient seen by Dr. Craig.
No complaints of racing thoughts or "abnormal thoughts".
She feels good in hospital.
Sleep good, appetite good, interest and energy good.

Plan:
Increase the Lithium to 300milligrams, by mouth three times a day, levels 0.35
Change Haldol to 4 milligrams, by mouth, every night as of tomorrow.
Discontinue close observation

Friday, December 1, 1995

I woke up at eight and took my bath before breakfast. I enjoyed breakfast and then went up to the bake and craft sale that our operating room nurses' association is hosting, but did not buy anything. I visited Wendy, the nurse who is relieving me, in my old office. It was nice to be there again if only for a few minutes.

I went to the group discussion on anxiety, and afterwards had lunch, still feeling very sleepy. After lunch, I tool a nap and then rewrote the Letter to the Editor on Health Care Reform. I went to a service in the chapel on AIDS.

After supper, I spent the evening playing pool, ping pong, and cards, after listening to the news. I went to bed early and slept well until I was awakened by the nurses making rounds, checking if I were asleep.

Clinical Clerk's Note
December 1, 1995 1450

Re: Altered thought Processes

Patient continues to deny having any weird or racing thoughts. Feeling much better since being in hospital. Complained of being tired and feels this may be due to Haldol and was wondering if it can be decreased.

Education provided concerning use of Haldol in treating bipolar illness, will need further education on bipolar illness. She did not take her Cogentin this morning-but wishes to take it this evening with her nighttime dose of Haldol (Cogentin is given to counteract the negative effects of Haldol like muscle twitching.)

Objectively she attends all unit activities. Hygiene is very good and appropriate. No pressure of speech and controlled facial expressions.

Saturday, December 2, 1995

I woke up at eight thirty and went to the dining room for breakfast. After breakfast P and I sat in the dining room wondering what we could do all day. We decided we would go into the craft room and make some Christmas decorations.

We made decorations until lunch time, then we spent time cleaning up the craft room. We could not find a broom, so I made one out of a crutch turned upside down with a towel taped around the arm piece. It worked very well.

Creativity and ingenuity should not be symptoms of mental illness. It could have been the first crude representation of a "swiffer"

Sunday, December 3, 1995

After breakfast we did some more crafts. I made twelve Christmas napkin holders. Eleanor came to visit me for a couple of hours.

Nurse's Progress Notes
December 3, 1995 0800-1600

General Condition. She is maintaining improvement. Tolerating medications, but taking Haldol at night 2 milligrams-not as prescribed. She spent a satisfactory weekend, socializing and interacting well with others. Patient is spending most of time working on crafts.

Monday, December 4, 1995

I spent quite a while doing Christmas crafts. I went up to the operating room and visited the girls. I am feeling a little tired.

(I went to an AA meeting with three of the other female patients.)

Dr. Pearce, the Defense's medical expert, thought it was unusual that I had gone to an AA meeting. I went not because I thought I was an alcoholic, but because it was out of interest and something different to do as I had never been to an AA meeting before. I have always liked doing different things. Once I had that psychiatric label, everything I did was considered abnormal.

Resident's note
December 4, 1995

50 year old female admitted three to four hours of feeling high. This was associated with increase in positive mood state.

Not psychotic at time of admission. She has longstanding history of bipolar disease. Some history of delusions of being "God"

Overall the patient has been improving consistently since admission

Lithium was increased from 300 milligrams every day to 300 milligrams twice a day and further increased to 300milligrams three times a day.

Prozac was discontinued on admission

Haldol has been changed from 2milligrams twice a day to 4 milligrams at night and Cogentin 1 milligram added.

No psychotic features, no racing thoughts. Speech more controlled and not pressures.

Overall improving.

Plan:

1. continue to decrease Haldol gradually
2. Continue Lithium at 300 milligrams, by mouth three times a day
3. Plan to discharge later this week.

Nurse's Progress Notes
December 4, 1995 1500

Altered thought process-continues to maintain improvement in this area. No psychotic features noted. No weird thoughts as on admission.

Re knowledge deficit: Ms Day verbalized that she does not like being on Lithium, however if it keeps her from "going psychotic" she will continue to take it.

Then she went on to say that she hoped after discharge she could decrease it to 2 lithium per day (total 600 milligrams.)

I described to her the relationship between serum lithium levels and the dose of Lithium. Instructed her regarding the therapeutic range of serum Lithium levels and the need to continue Lithium as directed.

She nodded in agreement, however did not verbalize any understanding.

I will continue to provide education re Bipolar illness and management. Given a pamphlet on Bipolar illness and will discuss same with her tomorrow.

She is hoping for discharge Wednesday.

Clinical Clerk's Note
December 4, 1995

Patient feels much better since admission, denies any abnormal thoughts or racing thoughts.

Expressed concern over:

1. The amount of Lithium she was taking- it was explained to her that the level in the blood was more important, not the total dose.
2. Feels Lithium is causing too much sedation
3. Lithium is causing decreased ability to concentrate
3. possibilities to return to work

She states mood is good – no depression. Appetite good. Sleeps too much.

Interest: Wants to start trying to get her ability to concentrate improved, discussing possible upgrading for employment.

Poor Insight. No DMV? Energy improved since admission, but still not at desirable level.

Patient only agrees to take 2 milligrams of Haldol every night and states will not take any more Lithium.

Lithium levels 0.58.

Plan;

1. Discuss with Dr. Craig re discharge. Patient requesting discharge Wednesday
2. Try to encourage and educate patient re Bipolar disease, however seems to lack insight.
3. Change Cogentin to evening dose.

Tuesday, December 5, 1995

I woke at 0830 and took a bath. I am feeling quite tired today. I did some more crafts. I phoned the nurses' union and was told that my appeal regarding my employment insurance will be heard around December 20th. Dr. Craig spent some time with me today and he is writing a letter for Blue Cross. I hope I will receive some financial compensation soon.

I had not received one cent of income now since May 15th as I was not eligible for employment insurance disability because I had taken educational leave the year before, and had not accumulated a sufficient number of week's work to qualify. I also had to be sick for one hundred and twenty one days before my long term disability insurance became effective; consequently, I also fell between the cracks financially.

We are decorating the lounge now for Christmas.

Rounds Note
December 5, 1995 1000

Patient is requesting discharge-still lack of insight. Perhaps discharge Monday.
Plan Increase Lithium 600 milligrams, by mouth, twice a day
Other medications remain unchanged.

Nurse's Progress Notes
December 5, 1995

Re: Altered thought Processes- continues to maintain improvement – Denies having any racing thoughts or "weird" thoughts. No pressure of speech, or inappropriate activity noted.

Re: Knowledge deficit of bipolar disorder- insight appears limited in this area. Education re bipolar disorder and Lithium is discussed in conversation, however she has not yet verbalized or demonstrated understanding of same.

Re: Tiredness- complaining of feeling tired. Napped in morning. She also complained of dry mouth, but no further problems.

Re: pass. She requested a pass to go out for lunch with friends. Two hour pass granted, to return by 1400 hours. However, she called at 1430 hours to say she was delayed. She has not returned yet.

Clinical Clerk's Note
December 5, 1995

Patient on day pass from 1200 to 1500, enjoyed this. –No complaints, no further thought disturbances, no racing thoughts, no pressured speech.

-Feels that change in Lithium is making her sleepy and is requesting decrease on Haldol.

Still poor cognition and lack of insight.

Plan:

Discuss with Dr. Craig re decrease Haldol

Blood work tomorrow.
Plan for weekend pass.

I was complaining of dry mouth, very tired and still Dr. Craig increased the Lithium amount to 1200 milligrams daily.

Wednesday, December 6, 1995

I did more crafts today. I went out today with the nursing students.
Blanche visited during the evening.

Thursday, December 7

I did more crafts. The nursing students had a Christmas Party.

I made enough crafts that week to complete my Christmas gift list with beautiful hand-made items that my family and friends really appreciated. I had learned a long time ago, it is not the cost of the gift that counts, but the thought and effort that go into it. I kept the napkin holders that I had made, and proudly display them every year on my Christmas dinner table.

I went out with Dad and Frances for lunch. I became nauseated and vomited. I have been feeling nauseated all evening.

Nurse's Progress Notes
December 7, 1995

Judi is doing well today. She agreed only to 1 milligram Haldol last night. Subsequently slept well and less drowsy this morning. Still taking Lithium. Plan for weekend pass with possible discharge Sunday. Will follow. Today's Lithium level still pending.

That was the last entry in my journal as I became too physically ill to write, with the increased dosage of Lithium that was toxic for me.

Off Service Note
December 8, 1995

Fifty year old female presented with bipolar disease mixed state. She felt high or hyper.
She has a psychiatric history of bipolar disease-mania resulting in being committed.
Since admission patient has done well. Her lithium dosage on admission was 300 milligrams, by mouth every day.
This has been increased to 600 milligrams, by mouth twice a day and yesterday Lithium levels are .77. Since admission her Haldol has been decreased. Patient has done well, no delusions/abnormal thoughts, no pressured speech.
Yesterday patient complained of feeling nauseated and vomited and still feels nauseated today, but no vomiting.

Nurse's Progress Note
December 8, 1995 1440

Judith has been nauseated and vomiting since 1100.
Dr. assessed her and told Judith to stay overnight and discharge tomorrow.
Judith pleased to stay as she does not feel up to going this afternoon.

2130 Re: vomiting

Patient continues to feel miserable.

She had a couple of episodes of vomiting early in the evening. None since around 1830 hours. She has been sleeping since that time.

Nurse's Progress Notes
December 9, 1995, 1030

Ms. Day continues to feel unwell. Energy is low and she has been vomiting this morning. No breakfast taken, just a small amount of Ginger ale. This has been kept down.

Dr. Craig notified and discharge to be held and reassessed in morning. Ms. Day resting on her bed

1410: Ms Day has been resting on her bed all day. She vomited prior to lunch, but she kept down the clear fluids she had at lunchtime. Vitals stable at this time.

Ms. Day states she feels like "I have the stomach flu" Continues to be on her bed all evening.

2130: Continues to feel miserable and vomiting several times this evening. She is unable to keep fluids and solids down. Lying in her bed all evening

2300: She did not wish to take her Lithium at bedtime due to her vomiting, therefore Lithium held.

Nurse's Progress Notes
December 10, 1995 0600

Judi was asleep in bed at change of shift, but she spent a restless night. She was awake times four. She continues to feel miserable with complaints of nausea, but no emesis.

She is asleep at present.

Resident's Note
December 10, 1995

Judi has continued to have nausea and vomiting since Friday. Refused lithium last evening. She feels the dose of 600 milligrams, twice a day is the cause – possible at present. No other problems otherwise.

Will hold Lithium for twenty four hours to see if this settles the vomiting.

Will reassess in the morning.

Nurse's Progress Note
December 10, 1995 0800-1520

Judi has been lying on her bed throughout the day. Lithium held this morning. She has kept down clear fluids and an apple with no vomiting. She said "it is strange that when I get to that dose of Lithium, I vomit." She now feels she is on the road to recovery. Vitals as charted.

2045

She has several episodes of emesis (vomited) this evening with ongoing nausea. Lying in bed all evening and unable to keeps liquids down.

Given sublingual Brozipam 0.5 milligrams at eight thirty.

To be nothing by mouth until assessed in am. Aware and agrees with this.

Resident's Note
December 11, 1995

Judi continues to have trouble with vomiting and nausea. Patient is convinced that it's all due to Lithium. As such, she refuses to take her full dose of Lithium.

Also very unwilling to go home.

Patient is unrealistic and demanding.

Refuses discharge, refuses medications- will discuss with Dr. Craig, question sensible to switch to Tegretol. (another type of drug used to treat bipolar disorder)

Nurse's Progress Note
December 11, 1995 1525 hours

Judi has been in her room all day. She stated that she vomited times one (not seen). Able to tolerate Ginger Ale this p.m. She has appeared in her night attire all day and still appears miserable looking.

Lithium reordered today.

Discharge order has been written.

Nursing Update
December 11, 1995 2200 hours

Judi continues to feel miserable this evening and doesn't feel well enough for discharge.

Her father will be in for her in the a.m. and she will go out with him.

Appointment will be set up with Dr. Craig for December 20th.

She continues to have emesis and refuses to eat her supper. On her bed all evening.

Resident Note
December 12, 1995

Patient discussed in rounds. Still planning for discharge. Lithium 0.2, not Lithium toxic. Advised to continue Lithium at a minimum of 900 milligrams a day – ideal dose would be 1200 milligrams, total daily. Vomiting felt not to be related to Lithium. No evidence of organic illness. No abdominal tenderness.

Plan: discharge.

Recommended minimum of 300 milligrams, by mouth three times a day.

Follow up December 20th.

Nursing Update re Discharge
December 12, 1995

Judi was called for breakfast, but didn't get up for same. Patient discussed in rounds. (see previous note).

Patient discharged home accompanied by father at approximately 1045 hours.

This is when I believe nurses have to be more assertive and as professionals, they must be better patient advocates. When they see a patient as ill as I was for the past three days, they should have insisted that something was seriously wrong, that I should not be discharged, and my symptoms and medical history should have been investigated further for other underlying illnesses. I could have been diagnosed properly two years earlier and that would have avoided all the grief that happened to my family and me for another two years.

Dr. Craig had actually written on his consultation record.

She is pushing to drop the haloperidol (Haldol) and in practice is taking two milligrams, HS, which is probably all that she needs, However, we are much better leaving the battle line at 4 milligrams, understanding that she will take two, then having it reduced to zero, in which case we will have difficulty in getting her to take the Lithium. The same logic will apply to continuing her Benztropine. (a drug used for the treatment of all etiologic groups of parkinsonism and drug-induced extrapyramidal reactions like muscle twitching, shaking and other weird contractures. This is scary!)

Again, a misdiagnosis in itself is not considered negligence, but the negligence kept occurring because nobody revisited the idea that perhaps they could have been mistaken or that I was suffering from symptoms that could have been caused by other illnesses. That possibility was not considered or investigated. This is where the negligence kept occurring. Due to this inaction, I was permanently injured.

I was being chemically lobotomized. Dr. Craig was ordering twice the dosage, to ensure that I would be getting what he thought I needed, due to low lithium blood levels.

I don't know what would have happened to me if I had taken the entire amount. I knew I must have been dehydrated from being nauseated and vomiting for so long as I had vomited everything I had drunk for a day or so at that time. I was so sick I could not stand up straight and was not allowed to stay in bed. I had to force myself to get up, get dressed, and leave the hospital with the help of my father. I was discharged very ill, and the physicians considered the vomiting was again psychogenic as "according to my blood levels," I was not toxic, when indeed I was.

Patient Review, Dr. Craig
December 13, 1995

Last week Judi was stable on Lithium Carbonate 600mg. bid which gave her a level of 0.77, which is in the low therapeutic range.

During an interview, she warned us that when she was discharged from the Waterford Hospital, she developed vomiting for several days and she characterized this vomiting as psychogenic in reaction to the stress of being in Waterford Hospital.

The plan for which she agreed, was for a weekend pass to start on Friday and all going well, to be discharged home.

Sure enough, Friday morning she developed persistent vomiting. Her Lithium levels as of yesterday were 0.2 so she is clearly not Lithium toxic.

She has been examined several times by Dr. Jones-Hiscock and shows no signs of any viral or other acute illnesses. This is clearly psychogenic in nature. She is also very resistant to the idea of taking any medication which, again, is true to form for the last fifteen years.

On the balance, I feel that this is likely to happen no matter what medication she takes and no matter when she is discharged from hospital and the only sensible recourse is to proceed with discharge as planned.

She has been advised to continue taking Lithium Carbonate although she has insisted that she is not going to take 1200 mg, per day and we have compromised by saying that 900mg. is an absolute minimum, seeing as her level at that dose was 0.58, which is within the maintenance, but not acute treatment range.

She has an appointment to see me in follow up on Wednesday, December 20, 1995.

Dr. Craig also wrote when he answered interrogatories for the medical malpractice lawsuit.

I say drug induced psychosis can present and mimic many psychiatric disorders, however, it occurs shortly following ingestion of the drug and resolves spontaneously when the drug is eliminated from the body.

Furthermore, symptoms are usually atypical of functional psychosis such as manic depressive illness.

In Ms. Day's case, there was no evidence of ingestion of a putative, causative agent, no rapid recovery. There was, however, a presentation typical of mania with psychotic features.

I have no personal recollection of any statements by Ms. Day concerning her previous psychiatric history or any disagreement with the interpretation of the history of her illness.

However I do have recollection, I was of the opinion Ms. Day did not have insight into her illness.

I say in my recollection, Ms. Day was accepting of the diagnosis of bipolar disorder and its treatment with Lithium until her visit of July 16, 1996.

I was a nurse educator and was well versed in mental illness. I had never been diagnosed with major depression, mostly anxiety and fatigue. I had denied being depressed many times and that was recorded on my medical records.

I had never been told I had been diagnosed with bipolar disorder until May 15, 1995. I knew I had never experienced any symptoms to meet the criteria to be diagnosed with that serious mental illness and kept insisting that my symptoms were drug induced, but none of the professionals took what I was saying seriously.

The physicians know very well about "substance induced mood disorders" as it is a very common event in hospitals for a patient to experience drug induced psychosis caused by drug interactions and overdoses, especially with patients recovering from accidents, burns, major surgeries, and so forth. The condition is called ICU psychosis. The psychosis normally disappears once the patients recover from the toxic drug assault on their bodies.

It clearly states in the DSM-1V-TR Diagnostic Criteria manual, which was in use at that time,[1] "manic like episodes that are clearly caused by somatic antidepressant treatment, (eg. Medication, electroconvulsive therapy, or light therapy should not count toward a diagnosis of bipolar 1 disorder." I was misdiagnosed with bipolar disorder.

Never go to one whose office plants look sick../Smallwood

Health Science Centre Psychiatric unit, 1st floor overlooking the pond

---

[1] Diagnostic Criteria from DSM-1V-TR pg.172

Chapter 13

Struggling while taking Lithium
December 19, 1995 to August 1996

Looking pale and feeling very weak, exhausted and nauseated, my father wrapped me in a blanket, laid me in the back seat of his car as we drove to Sibley's Cove. I slept most of the way back to my home town, but my Dad must have been experiencing a lot of worry, thinking about what he was going to do with me once we arrived there.

I could not eat anything for over a week and kept vomiting all fluids that I would force myself to drink. I spent most of the days in bed. Then one day, my father cooked some really salty turbot that was pure white in color, and brought it up to my bedroom. I forced myself to taste it and could tolerate eating it. It was the first bite of food I could manage to eat, and gradually my appetite returned.

I reduced the Lithium to half and kept on that dosage for a while longer, as I was still trying to believe that the doctors knew what they were doing and that I really needed to take Lithium to treat bipolar disorder, even though I could not tolerate it at such a large dose as the physicians had prescribed for me.

I missed the December 20, 1995 appointment to see Dr. Craig as I was too ill to consider riding in a car to St. John's, which was over two hundred kilometers away. The motion of driving in a car was too much for me to consider as I was feeling so nauseated all the time.

Those weeks before Christmas are just blurred memories, as I spent most of the time in bed. I felt very fatigued and physically ill, trying to tolerate as much Lithium as I could without vomiting it up. I did not feel sad or depressed as I was fighting too hard to just feel well and be able to eat and drink enough to stay alive.

Of course, my family members did not understand what I was experiencing as they were all counseled on the importance of my taking medication to stay well, therefore I was being preached to constantly "take your medication." Or questioned: "Did you take your medication?" Now I was definitely being treated like a child, and there would be no family support unless I took my medication and followed the doctor's orders.

As the days came and went, Christmas Day arrived. I had sent no cards to anyone, so I just gave my best friends and relatives the Christmas crafts that I had made at the Health Science Centre Psychiatric Unit. Both my children received nothing from me that year as I had not received any pay from May until December 22, 1995.

My children knew I loved them and they loved me. It definitely was a very hard year for them to experience their mother so ill. We always had had good Christmases together, exactly the same as when my brothers and sister were young. My mother and father certainly made sure we had everything they could possibly give us for Christmas.

Our relatives always sent us nice gifts too, so we had lots to open and our stockings were always full of fruit and candy. I have always loved Christmas, but this year I was too sick and tired to enjoy anything.

My brother, Jim, brought me his old computer to keep me occupied. When I felt better, I edited and typed my Mother's hand written Legacy for something to do to pass the time.

After Christmas, Dad and I drove into St. John's to visit our family there. I wanted to try to find another boarding house so that I could move back into the city. I saw an ad in the Telegram for a boarding house on Winter Avenue, and I called the number listed. When I heard the most caring, warm, and loving voice on the other end of the line, I knew this was going to be a good place to live.

I went immediately to look at the room and when I found out I could move in right away, I was very relieved and felt blessed to have found such a place, as hospitality, positive feelings, laughter, and love radiated everywhere, even from the pet dog Chrissie.

I did not think of how long I would be staying here as that did not matter as I only thought about today. By now, I had learned to live one day at a time, sometimes just one hour at a time.

The vomiting finally subsided, but the constant nausea and feeling of being unwell never completely went away. At first, I was barely existing and sleeping most of the time. I was fighting for my survival and to feel normal again.

I did very little housekeeping duties at my boarding house, as I did not have the energy to vacuum or do any other housekeeping chores, except to dry the dishes and put them away. I needed all the energy I had, just to keep my own room organized and tidy, plus performing my personal duties of daily living.

In the meantime, I felt it was very important to exercise my brain as well, so I had registered at Memorial University to study psychology 1000 and 1001. These courses were all memory work, and it took every waking hour to study the material, however I managed to get 70%.

I could not believe that my cognitive abilities had deteriorated to this compromised state and extent, in less than a year. I felt this unfortunate physical state I am in had to be drug induced as it had come on so suddenly, after being diagnosed with bipolar disorder and given psychotropic drugs.

I also took a philosophy course "Ethics of Modern Life" and wrote a paper on the wrong topic. I was surprised the professor gave me any marks at all. After that error, I phoned Professor Kiefte and explained to him that I was taking prescribed drugs and that my memory and cognitive abilities were greatly impaired. I told him I could not remember any of the theories and I was concerned about passing the final exam.

Professor Kiefte told me he understood, and that there was no way that I should fail the course as I had contributed so well in the oral presentations and discussions in class. I was reassured and felt better, as I could not bear the thought of failing a course, something that I had never done before in my life.

During one of those presentations in the Ethics class, my study group had to speak on the topic of Affirmative Action. I gave the presentation from a "Psychiatric" patient's perspective, without identifying myself as that person and asked them to imagine being locked up in a mental institution when you know you have never been diagnosed with a mental illness and you are not mentally ill. The more anxious and upset you became because you are locked up, the more the professionals attribute this to an assumed diagnosis of bipolar disorder. You try to explain to the staff that you are fine, but again they will not listen. The physicians keep on treating a diagnosis and ignoring you.

You ask your lawyer to help you get discharged and she advises you that the Mental Health Act overrules the Human Rights Act and you have no choices now. You must remain confined, as they are in control of you, and you must follow their regime.

The mental illness professionals convince your family, who have no background in medicine, that you have a chemical imbalance, and with the proper drugs, only a salt solution, you will live a so-called "normal" life.

While taking the drugs, you are so drowsy that you can hardly hold your head up. Your concentration, memory, and cognitive abilities are reduced considerably. You become more desperate and anxious, but still nobody listens. It is absolute torture. If you break down and cry, it is a sign of depression. If you awake at night with a racing heart and cannot sleep, you are manic.

The nurses and doctors interpret and record everything you say and do as being absurd. You have no idea that this is being done! You do not even have a chance to explain anything at the group meetings (rounds), and your appeal for release is ignored. You and your body are locked into a psychiatric trap! You are forced to take psychotropic medications that are monitored by blood level testing. It will be very difficult to escape from this entrapment!

I received high marks for that presentation, but I barely passed the course of Philosophy. At least I passed. My Grade point average has dropped considerably over the past year, but I still passed. I learned so much from this course, but most of all I understood more clearly the importance of affirmative action, particularly the importance of speaking up and out when there is a need to do so.

Dr. Craig's letter to Blue Cross
April 30, 1996

Further to our conversation of Friday, April 26, 1996, this is to confirm that while Ms. Day is much better than she was previously. I suspect that a return to her previous occupation of nursing instructor for the operating room theatres of the Health Sciences Centre is unrealistic at this time and that consideration should be given to rehabilitating Ms. Day for other employment.

Briefly, Ms. Day suffers from Manic –depressive illness. While I was not her treating physician at the time, I understand that she became acutely manic in May, 1995, was hospitalized for several weeks and then remained an out-patient for several months until she came in under my care sometime during the fall of 1995 when she was again manic. She is responding to Lithium Carbonate 600 mg. daily. While this dose is low, she is reluctant to take any more Lithium because of what she described as excessive fatigue. Fortunately, her Lithium levels are in the acceptable therapeutic range and her functioning is improving.

My functioning was not improving; it was getting worse. I really was not able to consider going back to work as the Lithium that I was consuming was making me ill. I was complaining of bilateral leg swelling and puffiness around my eyes. On May 29, 1996, I was referred to an internist, Dr. Amy Tong for a medical assessment. I was diagnosed with hypothyroidism, which was Lithium related, and I was started on Eltroxin, 50 mcgs. The Lithium was not discontinued at that time, so I kept taking as much as I could tolerate without causing me to vomit.

During the summer, I went back to my home on Inglis Place that had been leased, and spent time attending my garden that had been neglected for the past two summers. It was during that time when I noticed I was retaining more fluid in my legs, up to my knees, and also in my hands and face, especially after I was active. At this time, I was also noticing that I was experiencing more shortness of breath with any activity.

Before my illness, I had been in excellent shape, and now it was becoming difficult to walk up a flight of stairs. I had slowed down considerably. Despite trying to eat right and exercise, I now weighed over one hundred and seventy pounds. Previously I had normally weighed between one hundred and thirty-forty pounds.

That summer, I also attended meetings of the Consumer's Health Awareness Network of Newfoundland and Labrador (CHANNAL) in order to become more involved with mental health issues and communicate with other consumers. I wanted to have more understanding of bipolar disorder so that I could deal with this diagnosis and treatment as I really could not comprehend why I had to be medicated like this, as I had functioned so well all my life before being diagnosed with bipolar, manic in 1995. Since I have been treated with Lithium and other psychiatric drugs, I am now an invalid.

At one of those meetings, I met Roger who has bipolar disorder, and I told him what had happened to me in May 1995. I made the remark "the time was endless while I was locked up in the Waterford." He said to me "Judi, if you were really manic/ psychotic for those five weeks, the time would have flown so fast, you cannot imagine." I really knew by then I must have been misdiagnosed.

That summer, I kept studying, gardening, walking, and riding my bike while existing in one room in a fog of Lithium. For me, it was like pushing my car in neutral along the road, instead of being able to drive it. Life had become a real struggle. I was gaining weight without overeating, and even started to develop a fat pad at the base of my neck. I thought something was becoming terribly wrong, and I began to be very concerned. I was so drugged and profoundly fatigued that I did not do anything about it at that time as I was becoming even more physically ill and stupid.

Then one day, in late summer of 1996, I discovered that I was having considerable swelling, especially in my face, around my eyes, and in my hands and feet. My skin had become so taut; I could hardly bend my knees or my fingers. My eyes "squeaked" when I would rub them, like they were floating in a sac of fluid. I was also having difficulty breathing, even wheezing, despite using an inhaler.

I knew these symptoms were getting serious, so I made another appointment to see Dr. Craig as I felt I could not wait until the scheduled one. I sat in his office and said "Dr. Craig, this is not manic depression, I am slowly dying." "I cannot go on living this way." "Something has to be seriously wrong."

He examined my feet and legs. By this time the swelling was up to my knees, and my legs were the same circumference from top to bottom. He listened to my chest and sent me immediately for blood work, a chest X-ray, and an electrocardiogram, (ECG).

At this point, I had no idea that my ECG had showed "ST abnormality, Possible Digitalis effect" since I had been admitted to the Health Science Centre, in November, 1995, but this fact was ignored during that admission.[1] The doctors should have clued in by then that Lithium was affecting my heart, even though this is a rare side effect of Lithium.

I was going into congestive heart failure, as I did not have the ability to exercise at all without getting very short of breath and wheezing. You may say to yourself now: "She was a nurse." "She should not have let herself get this ill without doing something about it." The only answer I can give you is this: "These psychotropic drugs reduced me to a Zombie." At this time, my thyroid gland was dangerously being negatively affected by Lithium. I had no idea I was hypothyroid since the spring of 1995, and, I was suffering from fm/cfs/mcs. You could now say "she was reduced to the Village Idiot."

After my tests were completed as an emergency the same day, I immediately went again to see Dr. Amy Tong, who listened while I told her with tears streaming down my face, how I was feeling. I said that I was very sick, and that my whole existence at that time was fighting to stay alive.

I was physically weak. I felt very tired, always nauseated and dizzy. I had blurred vision, constant ringing in my ears, and problems with my balance, which was worsening. I was afraid that I might have a brain tumor. I was also having symptoms of Cushing's syndrome, as I was getting a fat pad at the back of my neck. I was experiencing severe shortness of breath on exertion, as if I were not getting enough oxygen, and I was breaking out in sweats. I also noticed that my hair was getting curly and frizzy. My hair had always been very straight and fine.

After the thyroid blood results were available, which took awhile, they were also abnormal and showed thyroid antibodies. Dr. Tong diagnosed my condition as Lithium induced toxicity and recommended the Lithium be discontinued. I still do not know how my chest X-ray and EKG turned out, as I could not find the results on my medical records when I obtained them in 1998. However, I am positively sure that I was into or nearing congestive heart failure at that time, as my body was full of interstitial (in the tissues) fluid, and, I had so much difficulty doing anything, especially breathing.

During my lifetime, I had never been considered a person who would be having psychosomatic complaints or being a hypochondriac. When I went to a physician and complained of something, there was usually a physical reason for the actual complaint, and medical treatment was provided. Dr. Wayne Button, my family doctor since 1980, stated that fact in court when he was testifying on my behalf after I was diagnosed with fm/cfs/mcs.

It was shortly after those visits to Dr. Amy Tong, that Dr. Craig reluctantly discontinued the Lithium. Dr. Tong ordered Diazide, (diuretic for fluid retention) 1 tablet a day for control. Dr. Tong advised me to stop taking Eltroxin and have my thyroid function test followed up.

---

[1] Joseph, Mary. M.D. and Vieweg, Victor. M.D. Electrocardiographic changes of sinus bradycardia and sinus Node Dysfunction among patients with therapeutic levels of Lithium. Depression 2:26-231 (1994/1995

Dr. Craig then started me on Tegretol, an anti-convulsant used for symptomatic relief, according to[1] CPS, from trigeminal neuralgia and mania. Dr Craig told me that I needed to take this drug because Lithium is known to reduce the seizure threshold and I would be in danger of developing seizures, plus it would also treat my bipolar disorder.

After I took Tegretol, it completely wiped out my memory and sense of being. I was absolutely in a stupor and couldn't even move my eyeballs. They felt dilated and fixed. I could hardly focus to see at all now. I was spaced out and floating in the clouds. I was doing very little of anything, so my days were spent at home resting.

One day, while I was taking Tegretol, the girls with whom I was boarding, noticed that I was becoming more inactive, so they decided they should take me for a walk. I became so weak that I did not think that I was going to make it back home as I could not lift my legs to walk. By now, I was beginning to shuffle instead of walk. My feet were apart for balance while I was dragging my left leg more than my right one. The girls stood on each side of me, held my arms and helped me walk back to the car, and drove me home.

After that day in August, I decided that I was going to reduce the Tegretol, as I could not function at all while taking that drug. I would try to gradually get off it altogether, and take my chance on having a seizure, as I thought that things could not get any worse than they were already.

Perhaps if I had a seizure, then somebody would take me seriously and figure out what is wrong. Over the next two weeks, I split the pills in pieces and took halves, then quarters, and finally discontinued them altogether. After just two weeks, I began to feel much better, even though I was still dragging my left leg, and found it very difficult to go upstairs. I was pulling myself up with my arms as my "up and down" strength in my legs was limited. I felt my head was clearing more each day and I was beginning to feel more like the way I used to be. My enthusiasm for wanting to do things was increasing as my energy was improving.

Not long after that, my friend and I decided to go on a vacation, which meant driving across the island of Newfoundland, nearly a thousand kilometers, to Gross Morne National Park. I really enjoyed that week, despite my fatigue and muscle weakness. My friend jokingly warned her daughter, to be quiet and not to torment us while I was driving, as I had to concentrate or else I might drive the car off the road. That was the standing joke for the week.

My friend knew that I had discontinued my medication. I felt that she was quite confident in my driving ability. A few years previous, we had encountered a fallen moose near Clarenville, NL., during a ski trip, but I was able to stop the car before we did any damage, except to bend my license plate. I pride myself with good, safe driving habits, especially since I survived that accident in a tunnel in Montreal in 1976.

During my next visit to Dr. Craig, I told him that I had discontinued the drugs, that I felt much better without them, and that I wanted a chance to prove that I did not require them. I was still not convinced that I had manic depression, as I had never experienced mood swings. I also told Dr. Craig that if I become manic, psychotic, or depressed, for no reason while 'not' on drugs, then I will understand that I have an illness that needs to be treated with psychotropic drugs for the rest of my life, and that I will take them. Otherwise, I would not be taking any more drugs as I could not function while taking them as they were making me disabled.

I also informed Dr. Craig that I was planning to move back into my home, which I had rented for the past two years, and I wanted to try to get back to work a month or so after that.

Around this time, I had met Paul through a mutual friend. He needed a date to go to a function, and I needed an escort to go to one as well, so we decided to accommodate each other. We began to develop a friendship, and we started seeing more and more of each other, and enjoying each other's company throughout the autumn.

---

[1] Compendium of Pharmaceuticals and Specialities, CPS 2008, pg. 2250

Paul was friendly, enthusiastic, intelligent, professional, loved music and dancing, and had a good sense of humor. He was light-hearted, easy going and a very good father of two amazing teenagers, who were very well-adjusted. The three of them treated me with the utmost consideration and respect, and, I enjoyed every moment I spent with them.

I felt that I was ready to try to begin a new relationship. After all, it had been three years since my marriage breakup, and for the past year, all I had been doing is struggling to survive. I still had confidence in the institution of marriage, but it would take a very special person to meet my expectations. I would never settle for anything less than what I know I deserve. Unfortunately this relationship did not survive my illness. I still have this opinion and hopefully one day I will be in a happy marriage based on mutual love, respect and trust, with a deep appreciation for each other as we both grow older together.

The doctor must have at his command a ready wit, as dourness is repulsive both to the healthy and the sick. /Hippocrates

Sibley's cove, where my father fished and we as children, played on those rocks.

Chapter 14

A Six Month Battle to resume Employment
September 1996 to March 1997

I had moved back into my home in September, 1996, when it became vacant, and I rented the in-law suite that was in the basement to a friend of mine. I also took in some boarders, to supplement my income. I was feeling much better without taking the psychotropic medications, but I still became tired easily and required a lot of rest. I could certainly manage to take care of myself and also my home and garden. Now, I wanted to get back to work as I was feeling well enough to do so.

Dr. Amy Tong, who was an employee of the Health Care Corporation, wrote the following letter to Blue Cross, dated September 11, 1996, received September 16, 1996. And "Yes", this is all she wrote!

Enclosed is the information on Miss Judi Day.
I am seeing her for Idiopathetic, (unknown cause) edema and Lithium induced mild hypothyroidism. This has responded well to Diazide, one tablet per day and discontinuation of Lithium.
Her main problem that may impede her from returning to work is her psychiatric status. I understand Dr. S. Jain and Dr. D. Craig were her attending psychiatrists and they are in a better position in advising on the ease-back schedule.

I thought Dr. Craig was supporting me with my decision to resume employment as I looked and felt much better without taking any drugs. With another eighteen pounds eliminated from my body, and more energy returning each day, I wanted to get back to my job. I loved my work!

I also thought that Dr. Craig was in favor with my decision, when he asked me to make arrangements for a meeting with my employer regarding my ease-back into the work force.

I arranged a meeting with the management and supervisory personnel of my employer, The Health Science Centre, who was also Dr. Craig's employer. I obtained the shock of my life when Dr. Craig spoke to them. He said that I had "fired" him from being my doctor, (which was not true), and that he was no longer my treating physician, which was also not true. Dr. Craig wrote on his last consult "She will return if and when she feels I can be of help." At that point in time I certainly had not fired him as I had assumed he was helping me with my job situation.

During that meeting, he also said "Judi suffers from manic depression and she requires drug therapy, which she is refusing. I cannot clear her for employment, if she does not follow my advice and take the drugs she needs." He excused himself shortly after that, and stood up to go. Before he left the meeting, he said even though 25%-30% of psychiatric diagnosis are inaccurate, he felt that he was correct with this diagnosis and that was it. He left the meeting.

I was left there in front of three senior staff members in a complete state of shock. I stated "if I have to be drugged, then I cannot function at the level required to perform my duties as clinical educator in the operating room." I knew that I was definitely in a trap, but I would find a way out of it. I said that I would obtain independent psychiatric assessments, and hopefully, get permission from another phychiatrist to return to work without being drugged, as I had done all my life before May 15, 1995. These supervisory staff members knew or should have known that I had done excellent work for the past twelve years.

This is the letter that Dr. David Craig wrote to Blue Cross, Dated October 8, 1996 and received by Blue Cross October 28, 1996.

I apologize for the delay in replying to your letter of August 22, 1996.

As we discussed by telephone on October 7, 1996, Ms Day is a 50 year old divorced nurse with an approximately 15 year history of manic depressive psychosis.

Unfortunately she has very little insight into her condition, insisting instead that she suffers from periodic psychotic reactions due to her hypersensitivity to a variety of drugs.

She acknowledges that I and at least one and probably several previous psychiatrists have informed her of our opinion that the underlying diagnosis is manic depressive psychosis, but insists that her first psychiatrist, Dr. S. C. Jain, informed her fifteen years ago that she did not suffer from manic depressive psychosis, but from this unusual sensitivity.

In any event, Ms. Day has become less and less compliant with the treatment and medication and most recently stopped Carbamazepine (Tegretol) sometime within the last month or so. She has no intention of taking any Carbamazepine, in the near future. She insists that she is ready to resume work without difficulty.

When last seen on September 25, 1996, Ms Day was somewhat more talkative than usual and mildly irritable, but was not either obviously depressed or obviously elated. Mental status examination was otherwise unremarkable except for her lack of insight.

As we discussed, I am in a quandary. While I am far from certain that Ms. Day is well, given that she is not prepared to follow my advice, the question of her resuming work will have to be resolved between herself and her employer.

I am sorry that matters have reached this situation, but hope that I will be able to be of some help in the future.

In pursuit of my being able to return to my job, on October 26, 1997, I decided to make application to the Waterford Hospital to obtain my medical records. Through this action, I hoped that I could obtain some understanding of how the diagnosis of bipolar disorder had been established, as I did not feel that I had mood swings. I knew that I definitely could not function while taking medication that were required to treat that disorder.

I completed the application to obtain my records on October 29, 1996. I had written on the letter that this request is made for the purpose of "reassessing the patient's diagnosis with regard to treatment and possible return to the work force."

Also in the pursuit of my returning to my job, I had two more psychiatric assessments completed, paid for by my long term disability provider, Blue Cross, and both agreed that I could return to my workplace without having to take drug therapy. These medical expert reports were done by Dr. Martin Hogan and Dr. Michael Nurse.

Christmas came and went. Of course I had the house all decorated as Karen was coming home from Fredericton for the holidays. I did not have the energy to do this myself, so I hired somebody to help me. I noticed that I could not hold my arms up long enough to put the decorations on the trees. My arms would become tired, and I would be in pain.

Paul and I were becoming more of a couple, and enjoying the time we could spend together, as his two teenage children took a great deal of his time and energy. Things were going well with our relationship, despite the time and energy I was also spending trying to get back to my professional life.

Sometime in January 1997, I had received the medical records from the Waterford Hospital, and while I was reading through them I discovered the errors and omissions regarding the diagnosis of bipolar disorder. For the first time, I became aware of the inaccurate damaging information from collateral history taking that was written on the progress notes. These inaccuracies, I believe now came from my estranged second husband, who should never have been interviewed, caused a snowball effect of my being misdiagnosed and maltreated for almost two years, and has caused me considerable physical, emotional, social, professional stress, and personal damages, which possibly could never be repaired, but I had to try.

I consulted my lawyer again who thought I should file a complaint to the Newfoundland and Labrador Medical Board. By having my medical care investigated by other professionals, hopefully the psychiatric diagnosis of bipolar disorder could be erased from my medical records. This would allow me to return to my work, and I would no longer be required to take psychotropic drugs that have caused considerable physical disability for me.

On January 14, 1997, Dr. Martin W. Hogan wrote to Blue Cross, Life and Disability Claims the following:

I am writing to state that at the time I saw Ms. Day on November 25, 1995, it was my opinion that her psychiatric condition was stable and I saw no reason why she could not return to work on an ease back program.

If you require any further information, please don't hesitate to contact me.

January 13, 1997 Blue Cross wrote The General Hospital Corporation, Health Science Centre:

Further to our telephone conversation of January 13, 1997, in which we discussed an ease-back to work program for Ms. Day.

Medical documentation does not support total disability as defined by the group policy, under any occupation definition.

Blue Cross will be engaging a rehabilitation specialist to coordinate the ease-back to work program. We will require, on or before January 27, 1997, a commencement date for the ease-back program so the proper arrangements can be made.

As requested we are enclosing a copy of Dr. Martin Hogan's release.

Should you require further assistance, please do not hesitate to contact our office.

On February 1, 1997, upon the advice of my lawyer, I wrote Dr. Robert Young, President of the Newfoundland and Labrador Medical Board and complained about Dr. Karagianis' treatment of me. I stressed that he had not taken the time to listen to me, that he had treated a diagnosis, not the actual symptoms that I was experiencing. I also stated that I disagreed with his claims that I had no insight into my illness. It should again be emphasized that since I had not been told by anyone of my diagnosis of 'bipolar disorder" before that admission, that it would be impossible for me to be aware of my situation. During my forced confinement, I had exhibited a normal mental status in a day or so, as soon as the drugs had cleared from my system, except I lacked his insight into my having bipolar disorder.

I also criticized Dr. Karagianis for failing to consult Dr. Jain in Ontario, when according to the clinical clerk, that Dr. Jain's hand written notes were illegible and the previous diagnosis was uncertain. I also complained that Dr. Karagianis had too quickly accepted second-hand unsubstantiated information from a psychiatric nurse from the Health Science Centre, who gave incorrect information that I had previously been diagnosed with bipolar disorder.

I also informed Dr. Young that Dr Karagianis had ignored information from my son who thought I was better. Dr. Karagianis had also misinterpreted my daughter's phone conversations.

I wrote that Dr. Karagianis had recertified me two weeks later with insufficient grounds. He had forced drug therapy on me that I could not tolerate. He also had not monitored me properly when I was on Lithium and experiencing side effects. At that time he should have re-evaluated the diagnosis and treatment, especially related to my thyroid functioning.

On February 13, 1995, I wrote a letter of complaint to the Newfoundland and Labrador Medical Board against Dr. Coovadia of the Waterford Hospital, regarding his signing of recertification papers, emphasizing the fact that he had not personally interviewed or examined me before he completed those papers.

I also wrote a letter of complaint to the Newfoundland and Labrador Medical Board against Dr. David Craig for unprofessional conduct and breech of confidentiality regarding his behavior in front of the administrative staff of the Health Science Centre during the meeting, when I had been trying to get back to work and he led me to believe he was helping me to do that.

While I was waiting to obtain responses from the Medical Board, I received a letter from Blue Cross stating that after trying unsuccessfully for six months to get a date for an ease-back to work, and now being deemed medically fit to enter the work force, my insurance company had no choice but to discontinue my disability benefits as of March 31, 1997.

I was devastated! What would I do? With no income and possibly not a chance of getting my job back, I would quickly become ruined. If I couldn't get back to work at the Health Science Centre and have an income, where would I obtain another job, being in this situation, with a diagnosis of bipolar disorder, and the inability to tolerate psychotropic drugs? I felt that I was defeated in every way! I started to experience anxiety as I had never felt before. I thought now that I would be trapped in psychiatry for good.

Again I regained my composure, calmed myself down, dug my heels in and promised myself that I will survive as I have survived these past two years. I am strong. I am a survivor, not a victim and I will win.

I paid special attention when I read Stephen Covey's words from 7 Habits of Highly Effective People

> "Many people think if you are nice, you are not tough, but win/win is nice… and tough". It's twice as tough as win/lose. To go for win/win, you not only have to be nice, you have to be courageous. You not only have to be empathetic, you have to be confident. You not only have to be considerate and sensitive, you have to be brave. To do that, to achieve that balance between courage and consideration, is the essence of real maturity and fundamental to win/win."

One day I will be strong enough to win/win, but right now I am only strong enough to have climbed to the highest tree and realized that I am in the wrong jungle, and it is detrimental to my health. I will find my own way back quietly and slowly, to greener pastures and then I will "win/win". Now is not the right time, as it would be to the detriment of my health. Nothing is worth that.

I had been trying to get back to work now since September 1996, after I discontinued Tegretol. Now it was the middle of February, 1997, and despite visiting the operating room and keeping in contact with the nursing staff and management, hoping any day I would receive a date to return to my position, I had not been successful and I was still unemployed. During that six months waiting period, I also had two independent medical psychiatric assessments that I knew were positive and agreed with me that I was well enough to go back to work, but I was still waiting and waiting.

I started writing in my journal again because writing down my thoughts had really helped when I was locked up in the Waterford Hospital. Now, I felt I was locked up in a psychiatric trap again. This time I was outside the walls of an institution, but experiencing the same torture.

Sunday, February 16, 1997

I awoke at 0130 with severe palpitations, and calmed myself down, but the waves of panic came over me. I knew that I needed to talk to somebody, but as before when I was going through a severe crisis, I did not want to bother my family or friends.

I thought of my options and the Kirby House was my answer. The Kirby House is a home for battered women in St. John's, where the staff provide a twenty-four hour counseling service. I felt that I was being emotionally battered by psychiatrists and the administration of the Health Science Centre.

I phoned the Kirby House crisis line and spoke to a person, whom I did not know, nor did she know me, nor did she realize the help she was to me as I became more calm, thanks to her help.

After awhile, she had to answer another call. I waited on the line for her to return, but I guess the other call was just as, or probably more urgent than mine. After a few minutes of waiting for her to return, I hung up and thought I would call again later if I needed to talk to somebody.

I could not get back to sleep immediately, so I began to read the book called The Joy of Stress. *After about an hour, I fell into another deep sleep and awoke at 0500, again with severe palpitations.*

I controlled them after a few minutes, but there was a lot of stress I needed to release. I began pounding the bed and shaking my feet to get rid of it. Finally it was time to start my day. Why does everything seem to get worse at nighttime?

Tuesday, February 18, 1997

Between sleep and being awake I read the compelling book, <u>Midnight Angels Fantasy</u>. It kept compelling me to finish and yes, I overdid it. Reading a full book in one night is too much, but it has kept my mind off my troubles.

By the time I had finished, it was 0500 and I was upset, frightened, and experiencing flashbacks. I controlled my heartbeat, but I began to cry and needed somebody to talk to. I phoned my daughter Karen in Fredericton. I lost control for a second and started to chant "My God, My God, and then God, God, God." The chanting only lasted about a minute, but that was too long. Karen comforted me, and reassured me I would be alright; that she believed in me, and that she would not let me go into the Waterford Hospital again.

She phoned my friend Paul on her other phone line and kept talking to me until he arrived to be with me. I thought "smart kid, I had taught her well." By this time, I was feeling much better, but still felt a little frightened and anxious.

Paul stayed with me until it was time for him to go to work. My friend, Anne, then came and stayed with me. Anne and I talked about what had happened. I showed her the book and she read the last chapter. She then understood how it could frighten me. We made tea, laughed, and chatted about nursing issues, as we are both nurses.

Paul phoned saying he would be coming home from work soon, and invited me to go to lunch. I went upstairs and soaked in the bath. When I looked in the mirror, I had a rash all the way from my neck to my breasts. The stress of my uncertain future was very apparent on my face as well. I calmed myself down and began to decide what I would wear.

I was completely at ease and went downstairs to the sunroom and put on a relaxing (slow music) tape that Paul had made for me while I rested on the chaise lounge.

Paul arrived shortly after with a very worried look in his eyes, but I reassured him I was fine. We then left to go to the restaurant.

After lunch, Paul and I decided that we should visit Dr. Wayne Button, who happened to be the family physician for both of us. He had been my family doctor since I moved back to Newfoundland in 1980. During the visit, I told Dr. Button about the flash backs and anxiety that I have been experiencing over the past week, since I learned that I wouldn't be able to get my job back, and that Blue Cross has cut off my disability payments. He said "Judi, you are in a catch 22 situation."

It was the first time I had experienced anxiety to this extent in my life. I told him that I have had the ability to control myself, and normally try to get over the episodes without incident with the help of my family and friends. I reassured Dr. Button that when I became too overwhelmed, I would report that to him and see a psychiatrist immediately, but, for now I was coping.

We both knew that I had been under considerable stress the previous two years while I dealt with my health and my inability to work. We also agreed that I needed constant supervision, and that I would not drive my car for a few days, until this anxiety subsides. It was the first time that Dr. Button had ever seen me in a state like this, which I knew was brought on by being exposed to severe and prolonged stress by losing control of my independence and my ability to be productive and self-supporting.

I believe that Paul was more upset than I was when we left the doctor's office. I knew that he was also worried about his work. After we arrived back at my place, Paul made some calls to his office, and I made some phone calls to recruit people to stay with me that evening, so that Paul could go home, change, and go to his hockey game. I knew he needed a break from the stress, as I could see he was suffering just as much from this as I was.

I felt that I would overcome this situation and I would be fine, but it was harder for him to believe that. This was the first time he has seen me anxious like this in the five months that he and I dated, even though I have been open and honest with him about my past history with psychiatry and the admissions to hospitals. Seeing me so well for the past months, it was hard for him to understand this change. It was difficult for me too, as I had never experienced anxiety to the extent of causing palpitations before.

On February 19, 1997, Dr. Nurse, a psychiatrist who testified during my lawsuit, wrote on his report to Blue Cross.

This patient, in the past, has had considerable difficulty tolerating psychotropic medications. I do not feel that any patient should be forced to take medication if they are not in agreement with the decision.

If one were to assume that this patient had manic depressive illness, this patient went from approximately 1983 to 1995, without requiring a subsequent hospitalization.

At the time of my assessment, the patient did not have psychiatric symptoms for which I could recommend that she currently be on any psychotropic drugs.

In relationship to this patient's return to work, whether this patient returns to her previous employment is really an issue between the patient, the insurance company and the employer.

I can only state that, based upon my examination and the information available to me, this patient does not show evidence, at this point in time in which I am able to make a diagnosis that she currently has an active psychiatric illness. I feel this patient is currently able to reenter the work-force.

Wednesday, February 19, 1997

I woke up at 0030 with palpitations, and I was able to calm my heartbeat down by relaxing myself as my mother had taught me to do. On the way to the bathroom, I noticed Paul was in the Pink room next door, sound asleep as he was snoring loudly.

After I wrote in my journal, I thought that I would read the novel Loving *by Danielle Steele. Maybe that would relax me enough to go to sleep again. It was then 0237.*

After reading myself to sleep, I woke again at 0337. I was not quite as anxious, but, this time I laughed to myself while thinking that these attacks were becoming as regular as contractions during childbirth. Then I began to read again. At 0405, I was ready for sleep.

0450 I am awake. This time I feel the scratching of my throat, and the symptoms of a cold coming on. This upset me, as my other two episodes of supposed "mania" were during times when I had been physically ill with colds and flu-like symptoms, and taking an antinauseant, Gravol, with other over-the-counter antihistamines. Then the next time it happened, I was prescribed Trinalin, an antihistamine, antibiotics, plus I took Prozac.

I went to the bathroom to gargle some mouthwash to see if I could make my throat feel better. This was not the time for me to get the "flu" I thought, as I had been perfectly well all winter until now.

I read from 0503 to 0535, and fell back to sleep but woke again at 0546 am, with my heart racing. I reminded myself that I was in no harm, and how Mom had taught me to remain calm.

I read the beginning of the book, Conquering Life's Most Difficult Times *and I started to feel that I was losing control again. My feet were sweating, which they never do, and I began to cry.*

I cried, pounded on my bed, and fretted. I went to the bathroom to have a warm, soothing bath to help control myself. (This had always worked for me at the Waterford Hospital, when I felt overwhelmed.) It took about ten minutes before I calmed myself down.

I dried myself off, put on my flannelette pajamas, and found my soft warm slippers that Dad had given me, and put them on. The slippers reminded me of my mother and I began to cry again as now I felt I was becoming an invalid like she had been.

Paul had just gone off to work when Dad arrived. I knew that he would take care of me now just as he had taken care of Mom for four years, after her strokes. He had also taken care of me in December, 1995, when Dr. Craig discharged me from hospital so sick.

Dad looked very tired, pale, and sad. His friend, Frances, had just left that day for Florida; one of his friends, Stan King, had died the previous night, and he was recovering from the shock of having a car accident last week. Now he was thinking he had to take care of me today because I was ill. I couldn't do that to him. I have to take care of myself.

Thursday, February 20, 1997

0108, I am awake again and I feel calm. I have all the time in the world now to relax and get better again.

I must write Dr. Evens to tell him that I will be absent from class. No, I think I can go to class Friday, and do my course, Biological Psychology.

0116, I can't sleep! Was it Winston Churchill who said, "It is a waste of time to be lying in bed, trying to fall asleep?" I will read for awhile until I get tired.

0151, I have read thirty pages of the book Loving, *and now I am sleepy.*

0200, I just woke up again, but this time I am completely relaxed. I can't think of anything worthwhile to write. I am not sleepy enough to go to sleep, so I may as well read again.

Thursday, February 20, 1997

It is 2351, I just woke up but I am very calm.

I realize now that I don't want to go back to the Health Science Centre to work. It would be too stressful. I have tried so hard for the past six months to get back to work, and the administration has put so many obstacles in my way that I must give up the idea of working there, as how could it ever be a friendly environment there again?

You would think that hospital administrations would be more understanding and accommodating to employees who become ill, but the opposite seems to be true.

I am not the only person who found it hard to deal with the Health Science Centre, once there were medical issues.

There had been a head nurse who had worked in the operating room for decades. After a trip to Montreal to learn about the operation of new laser equipment that was being purchased, she suffered from stress induced anxiety and depression. She was off work for months recovering from her burn-out. When it was time for her to return to work, she felt she needed a less stressful job until it was time for her retirement. She told me the Health Science Centre would not accommodate her, so either she had to come back to her old position, or resign, which she did, losing her ability to retire with a full pension.

She had dedicated her life to her work. She was let go by uncaring management because she wanted to work in another area of peri-operative nursing that was less stressful. What a shame! She finally got another job at St. Clare's Hospital, and worked in the endoscopy unit until she could afford to retire. (She developed cancer after that and died in her sixties.) From my observations severe stress lowers the immune system, and that leads to cancer.

I have discovered also that women are too easily diagnosed with mental disorders, instead of looking for underlying physical causes. Another class-mate of mine from high-school was having severe headaches. She was diagnosed with bipolar disorder and was treated accordingly. She told me that she would go to Old Perlican Hospital screaming in pain in her head, and would receive an injection, but nothing helped her with the headaches. I found out from a nursing colleague of mine that the doctors were ordering saline injections. They were giving her placebos for her pain.

Because of her supposed psychiatric illness, she was not taken seriously either and suffered needlessly. She finally had the proper diagnostic testing done which showed a brain tumor. She died shortly after that. Stress kills and I will not allow it to kill me!

Back to my journal

Thursday, February 20, cont'd.

I must be suffering from post traumatic stress disorder now. God knows I have been under enough stress for the past two years. I wonder how long it will take to desensitize me, like classical conditioning or cognitive behavioral therapy. I have to be strong, believe in myself, and remember my mother's last words she said to me. "You will be okay." Now I will read again.

Friday, February 21, 1997

0253, I stopped reading and felt sleepy.

0321, I just awoke again, and this time I was not anxious nor did I have palpitations. I have been reflecting over the past few days and nights.

I thought "you know I think I can really understand insanity." I believe it is nature's way of blocking out scenes and situations that become too unbearable to tolerate. It is identical to becoming unconscious when you have a severe head injury, otherwise you would not be able to bear the pain.

I remember when my son Brian had a head injury from falling to a cement floor from a scaffold at work, and the pain he endured. I saw him suffer so much that, at times, I wished he were unconscious; so that he would not have to endure it. Just touching his hospital bed very gently would make his pain more severe. I remember that he screamed in pain one day when I accidently jolted his bed.

I could not understand that until I read Florence Nightingale's book <u>Notes on Nursing</u>*. She mentioned that in her writings. It stuck in my mind as I remembered how painful it was for Brian, and how intensely a mother feels every pain her children suffer. It's incredible!*

It is amazing how clear your mind becomes in the middle of the night when you are wide awake. That was when Brian did most of his studying. At a very young age, maybe fourteen, he asked me not to nag him to study any more as he could only study when "he" wanted to.

That made sense to me, so I did not remind him any more to study. I remember, though, when he worked so hard during grade twelve to get his scholarship. He would wake up in the middle of the night to study. Brian worked hard in school, and did well. I kept telling him how proud I was when I went to his teacher's interviews, and received so many good reports. I always told him how smart I thought he was, and I called him "Brain" instead of Brian. He was a genius when it came to emotional intelligence and really had the street smarts too. He does well with his business and works very hard, providing for his family.

I remember Karen's Grade 1 teacher saying to me during an interview "If I could chose one of the children in my class to be my daughter, I would chose Karen. I was so proud of her too as she always excelled in everything she did as well in radio, television and running her own business until she retired with the luxury of being a stay-at-home mom. My children were my whole world and nobody came before them.

Saturday, February 22

It is 0119 and I am awake, but not anxious, just calm. I read the paper until I was sleepy again.

It is 0328, I just woke up again and I am still not anxious. I do not have palpitations and I began to think about myself and how desperate I have become.

Yes I will write my book. This book will be to Judi, for Judi, from Judi, and try to tell the story about the desperate person I have become, trying to find answers and how deep I had to dig within myself to find the facts about living.

It is 0434, I just woke up again and still I am not anxious. I feel relaxed, but tired. I feel too tired to go downstairs to get more paper.

I woke at 0630 and again I was serene, but I started thinking about Dr. Doucet, the psychiatrist Dr. Button found for me, who he felt was understanding and objective, and would possibly help me so I may return to work. I was wondering what he will tell me as I believe I would have to listen to his advice

Will Dr. Doucet read this mess that I am writing first and then listen to me? I am tired of talking and nobody listening to what I am trying to tell them. The experts write it in my charts as "pressured speech" but it is caused by being so frustrated with the situation that I am in. Then I thought of Dr. Jain and imagined seeing him behind the desk. I became very weak, sweaty with palpitations and began losing control. I took a relaxing bath and finally settled back to sleep.

I'm going to get dressed now to go to the funeral home to visit Stan King's family with Dad as I wanted to see Mom's good friend, Ethel, and give her my condolences.

After we went to the funeral home, Dad and I left for Sibley's Cove with me driving his car. I felt fine and we had an enjoyable trip, stopping in a store on the way to get some food. Paul drove out separately in his own car, and I had plans to drive back with him on Sunday.

After dinner, my brother Max and his wife Nita came for a visit and Dad opened a bottle of home made wine. That was where I made another mistake. I drank a glass of this old red wine. Since I had not drunk any alcohol for a long time, it did not agree with me.

I went to bed, as I was tired, and went to sleep immediately, but woke up in the middle of the night very anxious and upset. I remember Paul trying to hold my hands and I fought him until I came to my senses.

I got up then and went downstairs and was trying to keep myself busy doing things until everybody else woke up. I polished Mom's silver spoons that looked like they hadn't been cleaned since she died and tidied up the living room and kitchen.

Since Mom died, Dad appreciated the little things that I would do for him around the house. Even though he kept everything clean, neat and tidy, there were always the little incidentals that needed the woman's touch.

Despite keeping busy and doing house chores, which was the only way I could think of to keep my mind off the horrible situation I was in, I felt very anxious and panic kept creeping back in.

Max and Nita visited after breakfast and Max and I went for a walk to see if that would calm me down. We walked and talked and when we arrived back, we decided to go down to Old Perlican Hospital to see if I could get some medication to help me relax as by now I thought I needed something, as I had reassured Dr. Button that I would seek medical help if I could not control the anxiety.

This is where another nightmare began. I was pacing the corridor until I was checked in. Then I was placed in a private examination room with a stretcher. I laid down on it while waiting for the doctor to come and see me.

I had no idea that Max and Nita had told Dr. Nell that I had bipolar disorder and that I had stopped taking my medication six months ago. Of course, Dr. Nell called St. Clare's Hospital and the psychiatrist on call was Dr. Karagianis, who should have known by now that the diagnosis of bipolar disorder was questionable, since he had received the complaints that I had written to the Medical Board. The last thing I should have been given was antipsychotic drugs, especially Haldol.

Despite Dr. Karagianis having been informed of a possible misdiagnosis by the Medical Board, he ordered Dr. Nell to inject me with Haldol 10 milligrams and Valium 5 milligrams. The next thing I knew Dr. Nell was standing over me with a needle.

I tried to talk to Dr. Nell, but he did not listen to me. I became more frustrated and started banging my hand on the side rail and begging him not to inject me with Haldol as it was a wrong drug for me and I did not need it. I was banging my hand so hard; the pearl I was wearing came out of its setting. I was really struggling to try to get somebody to listen to me and not inject me, but he ignored what I was trying to tell him.

Instead, without any more time wasted, I was held down and Dr. Nell forcefully injected me with those powerful neuroleptic drugs, against my will. I was transferred to a portable stretcher, and restrained there, placed in an ambulance to be driven to St. Clare's Mercy Hospital in St. John's, almost two hundred kilometers away.

While lying on the stretcher in the ambulance, I was becoming very nauseated as I get motion sick easily. Now, with these drugs in my body and lying down flat, I knew it would not be too long and I would be vomiting while lying on my back strapped to the stretcher.

I pleaded with the nurse, who knew me personally, to please remove the straps because if I start vomiting, she might not have time to get me on my side, and I might aspirate my stomach contents. I said that it will be too late once I start vomiting, and there was no doubt in my mind that she would have had a casualty on her hands.

Thank God she agreed with me, asserted herself, made the decision to remove the straps. This allowed me to sit up in the seat, so I could look straight ahead to ease the nausea. For the next three hours, I sat in that position, chatting with the attendants, until we arrived in the Emergency Department of St. Clare's Hospital.

This letter accompanied me:

Dr. Caragianis; (Karagianis)
Re Judi Day

Dear Dr. Caragianis (Karagianis)
Thank you for accepting this 51 year old nurse as discussed in acute manic episode.
Treatment given Haldol 10mg. IM
                    Valium 5 mg. IM (no Lorazepam available)
Please assess and treat accordingly.
Sincerely,
Gert Nell.

Dr. Karagianis ordered Dr. Nell to inject me with these heavy doses of unnecessary dangerous drugs. Dr. Nell denied in the courtroom that he had injected me against my will. I knew he did as I was well aware of what was happening to me. I will always remember that assault and battery and the torture I went through before, during, and after the injection of those dangerous drugs.

In September 2003, I sent Dr. Gert Nell a notice of expert witnesses, pursuant to rule 34.05

> Dr. Nell will give evidence that he treated my parents for many years and knew them socially, and neither one of them had any evidence of depression, or other mental illnesses.

> Dr. Nell will also give evidence relating to a visit I made to Old Perlican Hospital, February, 1997, which resulted in my being involuntarily injected with Haldol, despite adequate evidence that such a powerful toxic drug was required, and transferred me to St. Clare's Hospital upon the orders of Dr. Karagianis, who should have known at that time that the diagnosis of bipolar disorder was questionable.

> Dr Nell will also give evidence of my behavior and credibility when I cared for my father, suffering from bone cancer, for three months before his death in June, 2001.

> Dr. Nell wrote to his lawyer on this court document.

Paul: Even this letter does not make sense. (1) Basically to me she is saying there was evidence for injecting her with Haldol, which is exactly what I did.

During the medical malpractice lawsuit, Dr. Nell testified that he did not know who injected me. I showed this evidence to the judge as a lawyer had given me the document. The judge stated it was privileged information between Dr. Nell and his lawyer and he could not accept it as evidence. Dr. Nell had lied during my medical malpractice lawsuit and he has never been charged with that serious action of being deceitful.

Other physicians were untruthful as well, but it was my word against theirs so it could never be proven that they had actually lied when they were under oath to tell the truth and nothing but the truth.

Hippocrates Oath, First do no Harm.

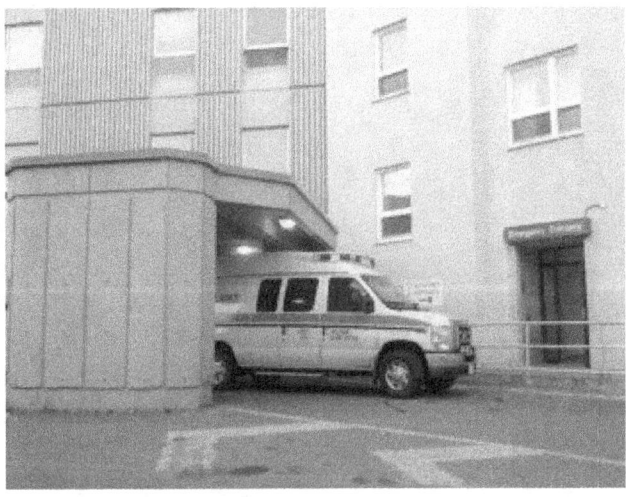

Emergency Entrance, St. Clare's Hospital where I was sent from Old Perlican Hospital.

Chapter 15

First Admission to St. Clare's Psychiatric Ward
Feb. 23, 1997-Mar. 3, 1997

Sunday, February 23, 1997

I am admitted to St. Clare's and I am very drowsy and sleepy. I ate meals, but do not remember much more.

Medical status exam
February 23, 1997

Fifty one year old lady, looks her stated age, well groomed, orientated times three, tired and anxious and irritable at times.
She is cooperative, good eye contact, pleasant.
Speech normal tone and volume. No pressure of speech
No delusions. No hallucinations.
Mood not depressed, affect reactive. No suicidal thoughts.
Attention and concentration good.
Her short and long-term is memory good.
Insight fair. Admitted because of plus, plus stress.

Chief complaint:

Increased anxiety and poor sleeping.

History of Present Illness:

51 divorced lady from St. John's with long history of psychiatric disorder since 1980s. Followed by Dr. Jain, Dr. Karagianis and Dr. Craig in the past.
No follow-up since August 1996. Her Lithium discontinued on August 1996.
She has an appointment to see Dr. Doucet, not sure when.
She was doing well until two weeks ago when she found out that her disability will be discontinued on March 31, 1997.
She states that she was trying to go back to work since September, 1996, but she is unable to do that and she doesn't know why.
She states that since five days ago she has had poor sleeping with middle insomnia.
Unable to sleep last night. She spent last night at her father's home.
According to her friend and her brother she became increasingly anxious during the last two days, walking around the house. Poor appetite.
No grandiose ideation. No pressure of speech.
According to her family, denies any suicidal ideation.

Feeling very anxious this morning and she was seen at the hospital in Old Perlican and she was certified with one signature and received Haldol 10 mgms i.m. and Valium 5 milligrams today and transferred to St. Clare's Hospital, emergency room for psychiatric assessment.

She denies any alcohol or drug abuse.

She refused the admission but after further explanation of her current situation, she agreed to stay at the hospital.

Physical exam was done and documented but nothing significant noted.

Impression:

Adjustment disorder, increased recent stressors,
Rule out bipolar disorder.

Plan: Discuss with Dr. Karagianis
Admit to 3 west for observation.
Ativan I milligram, sublingual when necessary three times a day.
Reassess tomorrow.
Close observation, routine blood work.

Nursing Assessment
February 23, 1997

Main source of information; male friend, brother and sister-in-law

Allergies:

Sensitive to codeine/morphine, Demerol antihistamines.

There is no mention on my medical record of my sensitivity to Gravol, Lithium, Prozac, Tegretol, benzodiazepines and cold/ flu medications. This is where erroneous medical histories let patients down. Medicine and nursing are equally responsible for these errors, but this careless history taking has to be addressed.

Family:

Mother died two years ago-history of depression

You may also note that the family history changes from no family history of mental illness when I was admitted in 1983 to positive family history of mental illness, that mother suffered from depression." My mother was never treated for major depression even though she survived cancer, diabetes, strokes, and loss of her eyesight. She was very positive, spiritual and was an inspiration to anybody who knew her. She had the same interests and activities that I have; gardening, writing, and listening to tapes of books, after her eye sight failed her.

When the hospital Chaplin came to visit my mother before one of her operations, he said to her. "Mrs. Sparkes, I came to console you before your surgery and I am leaving you now, being uplifted myself." My mother did not have the illness of major depression or a depressive personality, any more than I have. It bothers me that this has been recklessly recorded on my medical records. Because of erroneous charting, there is a family history of mother, father and brother recorded as having a mental illness.

Both my mother and father kept themselves active by travelling and socializing with their many friends. Mom loved to live each day to the fullest and was content. I cannot remember her being in a negative mood or ever lose interest in her life, despite her chronic illnesses.

Her medical records indicate that she was hospitalized in 1974 for assessment of anxiety, diabetes, and hypertension. She was actually very upset at that time, due to a crisis that had resulted from an issue in the church in Sibley's Cove, where she was ordered by the committee to publicly apologize in church to one of the members.

This man to whom she was supposed to have given an apology, had kept playing a scratched record over the loud speakers, which sounded terrible, so my mother walked up to the church to ask him to please remove the record and play another so that the chimes our family had donated, in memory of my Grandparents, would sound like they were supposed to sound.

My mother was so upset about being humiliated, as she felt she had done nothing wrong, she cried off and on for two days and ended up being hospitalized due to her hypertension and diabetes going out of control. It was recorded on her medical record "reactive depression".

Crying is a release and is healthy, if it can be stopped after two minutes, not two days. I had learned that fact during a psychology course and how to protect myself from the negative results of crying, I stop and use Cognitive Behavioral Therapy by thinking more pleasant thoughts, and doing interesting things. The technique is called "stop thought" and it works wonders.

The Minister denied the request by this man for a public apology from my mother, and life went on as usual in Sibley's Cove. Years later, this man accidently knocked my mother and another lady down by backing his car into them in the church parking lot. Fortunately, neither one of them was killed, although the other lady had her hip fractured and other serious injuries, my mother had no fractures. After a few weeks, she recovered fully from her soft tissue injuries.

This is an example of how easy it is to have a family record of mental illness that can negatively impact other members of the family when as a result, they are labeled and stigmatized.

Appearance:

Arrived in wheelchair, blonde hair, clear complexion, "green eyes", attired cream pants, cream and brown sweater

Please note that I arrived in a wheelchair as I was too weak and tired to walk due to being injected with massive amounts of Haldol and Valium against my will.

Other Stressors contributing to illness:

1. **Presently difficulty with review board not accepting letter of clearance to go back to work. Patient very stressed by same.**
2. **Patient has great difficulty accepting her diagnosis. She refused prescribed medications of Lithium and Prozac.**
3. **Poor insight into illness.**

Patient had gone home for weekend (Sibley's Cove) on Saturday. After having cooked supper, behavior change was noted in patient.

Patient began to exhibit some manic behavior at her father's house –changing pictures and rearranging furniture to the disdain of her father. Incorrect, my father always appreciated what I did to help him.

According to the patient she says she became anxious at 12:01- she then began having thoughts which caused her to become progressively more anxious.

Patient said she began to pound the wall, clap her hands, slap herself. This activity helped her relieve the anxiety.

Patient was able to be in bed, but did not sleep, She was up at 0550 rearranging ornaments – family were notified at 0800- by this time patient's behavior was described as "frightening" by family- cursing, shouting, pounding walls and saying "driving nails."

Patient was taken to Old Perlican Hospital, where she required restraining to have Haldol and Valium injections.

Patient was then taken by ambulance to St. John's.

Patient was able to answer questions about night prior to admission, but fell asleep during interview.

Remainder of information was obtained from male friend, brother and sister-in-law.

Mental Status:

Settled on admission.

Pleasant cooperative, orientated times three spheres. Speech normal rate and volume.

Risk for Falls:

Unsteady on feet.

Nurse's Progress Notes
February 23, 1997

51 year old female admitted to care of Dr. Karagianis to be transferred to Dr. Frecker in morning.

Patient was drowsy upon admission and slept before initial assessment could be completed.

Family gave history of patient's illness.

Patient has been under a great deal of stress this past year since being on long term disability.

Patient was given clearance to go to back to work, but review board rejected it for lack of information.

The doctor, who patient was seeing, was unable to complete required forms as he was in conflict of interest.

The patient is under stress because her disability is due to run out March 31, 1997.

Patient is the type of person who has to have everything under control and does everything quite promptly.

Patient's condition gradually deteriorated over the past two weeks until it deteriorated to the point where patient lost control.

This weekend her family described her overall behavior as 'frightening'.

Patient was loud, unable to sit still, cursing, banging the walls, clapping her hands, arranging furniture.

Patient's brother admitted that he has noticed her behavior is not right.

Patient refuses to admit anything is wrong and becomes upset at the mention of drugs Prozac and Lithium.

He noticed that she appeared bloated when on Lithium.

He said she is the type who will help anyone out, but cannot solve her problems.

He said she refuses to admit she has a bipolar illness, which she has been diagnosed at Health Science Centre and Waterford.

Her family described her as a strong independent person who has had a lot of stress in her life.

Patient still harbors a lot of resentment towards her admission to Waterford Hospital and has not resolved her relationship with her brother for his involvement there.

Brother stated that "patient had access to her chart at Waterford Hospital and has a copy of same in a file at home."

Patient's brother expressed concern that patient would gain access to her chart here and therefore jeopardize their relationship.

Patient's brother states that patient stopped Lithium and Prozac abruptly last August, 1996 and has only taken Premarin and vitamins since.

Since arrival to unit, patient has been laying in bed with episodic periods of clapping hands or banging side-rails- at one point patient expressed- "I'm sick, I'm mentally sick, no, I am not mentally sick."

Patient was given Ativan I milligram, sl at 2115.

Patient accepted this medication without difficulty.

Patient overall settled.

Patient maintained on close observation.

2300

Patient up to washroom but did not void. Settled overall.

Monday February 24, 1997

I slept pretty well.

1800, Max, Nita and Paul visited

It is now 2230, I have slept since 2130 and I was awakened by shoes clicking along the floor of the hallways.

I thought of Paul as soon as I awoke and wondered how he was and why he did not answer the phone when I called, or at least, left his message manager on so I could leave him a message that I was fine, and tell him that the nurse came in to see me after he left. I believe this was the beginning of the end of our relationship.

Clinical Clerk's Notes
February 24, 1997

51 year old separated lady from St. John's. Operating room nurse on disability pension since one year ago.

She stopped her Lithium on August, '96 because of side effects.

Previous admissions to Health Science Centre and Waterford Hospital in May, June 1995, and November 1995 and diagnosed with bipolar disorder.

She stopped the Lithium on August 1996. According to the patient, she is doing better since then.

She was trying to go back to work since September.

She saw Dr. Hogan for psychiatric assessment with regard to the diagnosis and he suggested that she can go back to work.

Also she was seen by Dr. Nurse for the psychiatric assessment.

Recent increased anxiety and admitted on February 23, 1997 for observation and further assessment.

Discussed at rounds today and assessed by Dr. Frecker.

Doing well since admission. No management problems.

Slept well last night. Complaining of feeling drowsy today because of the medications (10 milligrams Haldol, 5 milligrams Valium given at Old Perlican Hospital, and I milligram sublingual Ativan last night PRN.)

Denies any thought disorder.

No pressure of speech. cooperative and pleasant.

Mood not depressed, affect reactive.

No suicidal thoughts, attention and concentration good.

Memory long and short term intact, good insight, good judgment.

Impression:

Adjustment disorder with increased social and personal stressors.

Plan: Change her status to voluntary patient.
Observation.
Support therapy.
Lorazepam decreased 0.5 milligrams sublingual three times a day whenever necessary.

Nurse's Progress Notes
February 24, 1997, 0630

Client appeared to have slept well throughout the night.
Maintained on close observation.

0800-1600

She ate breakfast in her room. She is resting on bed after breakfast. Patient reports "feeling tired" this morning due to medication she "was given last night."
Patient not accepting of her illness, diagnosis, bipolar
Patient reports information about her that was provided by her family to the doctors was misinterpreted."
Briefly discussed work issues as she reports this "upsets" her. Support provided.
Denies any thought of suicide.

Discussed in Rounds:

Medication review with reorder done. Medication changes made.
Seen by Dr. Goughi and Dr. Frecker after rounds.
Status changed to voluntary.
Napped again after lunch. Up and about with encouragement.
Socializing later in afternoon.
Received phone calls from family and friends.
Close observation discontinued and routinely observed for remainder of shift.
Consult to occupational therapy sent.

February 24, 1997 2130

Patient's male friend reports patient much improved over yesterday. He said she is tired tonight because of lack of sleep and because of medications. Patient slept as soon as he left. Sociable during evening.

Tuesday, February 25, 1997

I have been asleep and have had several power naps so far tonight lasting about one half hour each. Now I am going to read for awhile.
0120, Sleepy again.
0145, Awake again. This time I was cold so I got up to pick my blanket off the floor.
0150, Asleep
0250, Awake again with a very dry mouth and my nose, and my sinuses were burning, I guess with the dry air in the building.
0300, Asleep.

0312, Awake again. The click of the nurse's shoes while walking woke me. It is amazing when you are a light sleeper what wakes you up! Nurses should have soft soles on their shoes when they are making rounds.

0315, Asleep.

0330, Awake again, becoming more alert as the drugs are wearing off. My head is becoming much clearer. I don't want to take anything else.

0335, Asleep.

0400, Awake again. No palpitations. Otherwise very calm.

0420, I am feeling fine, no anxiety whatsoever. I feel that I am getting all my frustrations out of my system, and I will be on the road to recovery, drug-free, I hope.

0430, Asleep

0450, Awake again. I am feeling very cold, so I went to the desk and asked for a blanket. There was none, so I took a couple of sheets to wrap around me and went back to bed

0455, asleep

0530, Awake again, feeling fine and warm. These blankets are really warm. I only wished more time had passed while I was asleep.

0545, Asleep

0600, Awake again, It must have been the snow plow that woke me, but I still have no anxiety or anything.

0605, Asleep.

0700, awake and took a bath to get ready for the day. I washed my hair and I did not feel as drugged. I am feeling good. I need to phone Nita regarding my check book.

0800, I ate breakfast

0900, I did woodworking at Occupational Therapy. I sanded my wind chimes and another patient sanded my CD and Tape stands.

1100, I had a visit with Dr. Goughi. Mary visited during the interview just for a few minutes. I had lunch, played cards and went back to occupational therapy and began painting a clown.

1630 to 1730, I Played cards.

It is 2400, I am awake again this time I feel cold and clammy.

Clinical clerk's Notes
February 25, 1997

I spoke with Ms. Day today. Doing better today. States she is feeling much better. No evidence of manic depression.

Plus, plus stress including social and personal during the last three years. She is separated since three years ago and off her job since two years ago.

She was trying to go back to work since September and difficulty with regard to her disability pension. She is not sure when she can go back to work.

Her mother died in June, 1995 because of cardio vascular accident while she was at the Waterford Hospital. She states she was very close to her mother.

Plan:

Observation
Supportive therapy
Possible discharge after weekend
Continue with PRN medications (whenever necessary)
Patient has appointment with Dr. Doucet on March 6.

During this admission, it is recorded that there was no evidence of manic depression.

Nurse's Progress Notes
February 25, 1997 0800-1600

Up and about unit. Affect bright. Pleasant upon approach. Reports feeling improved since admission.

Denies feeling anxious. Writing in Journal.

Patient reports she hasn't had her Premarin in past couple of days. Dr. Goughi aware and some ordered.

She attended occupational therapy and relaxation today. Enjoyed same. Routinely observed during the day. Socializing at times.

February 25, 1997 1600-2400

Patient overall settled in lounge mostly all evening lying on sofa appearing quite relaxed. Patient admits to feeling that she had a good day.

She asked whether there was any talk around Lithium.

Patient was encouraged to take it one day at a time and for now this drug is not considered,

Patient again reiterated that she had a lot of stressors in her life and feels that this is to blame for the way she is.

She said all the stress came together and she could not cope.

She is relieved now that she has some answers. She accepts she is not ready to return to work and that her disability will not be cut off.

Routinely observed.

Patient's boyfriend visited expressed concern re patient's behavior.

Said patient becomes upset at the mention of drugs- Friend took the wait and see approach, but patient, it seemed, jumped ahead and said that if he could not accept her without drugs then she would have to terminate the relationship.

Patient admitted to not being able to handle any type of stress and that she is quite sensitive overall.

Patient said she finds it difficult to handle conflict of any nature and becomes easily defensive.

Reassured and support given. Routinely observed.

Wednesday February 26, 1997

0150, I just woke up again. This time I am very warm, even perspiring, and now I am beginning to feel damp and cold. I am still very sleepy, so I will lie down again and sleep. I can barely keep my eyes open to write.

0200, I don't think I slept that episode. I was only thinking while awake. I thought how strong I am both mentally and physically. I have endured more physical and emotional pain than most others. I have many family and friends who support me, and I know that I will be okay

I don't think that I will have another panic attack or anxiety attack like I had, because I am not under the same pressure as before. I know now that I am presently not well enough to go to work, especially to the Health science Centre. Thankfully my long term disability insurance will not be cut off on March 31.

I feel secure now, both emotionally and financially, and everything else in my life is wonderful. I have the best friends and family, (my children, Karen and Brian), whom I love very much.

0315, back to sleep

0430, I just woke up again. This is the first time since I have been here that I feel anxious. It was like I was reliving the "Waterford Hospital" and I began to get afraid that the nurses would think that I was going "manic" like they did at the Waterford Hospital.

I am perfectly calm. I am just having a little difficulty staying asleep more than an hour.

0435, Back to sleep.

0455, Awake again. This time I am having palpitations- I don't know what I am getting anxious about now. I have asked for an Ativan to help me relax and go back to sleep.

0500, I am going to get back in bed and hopefully sleep until 0700.

0515, I am wide awake again as calm as a cucumber. The first thing that came in my mind was the play Macbeth, where we are on stage and each of us is playing the part. I became wide awake and worried about what is going to happen to me next. I stopped that and decided to get through the rest of the night uneventfully.

0550, Someone was talking very loud outside my door and it woke me up. I am not anxious or anything, but I cannot lie in bed once I am awake. I put on my housecoat as I had chills.

0555, I am sleepy again so I will try to go back to sleep.

0600, No sleep will come- I thought of my sister-in-law, Nita and how good she was to my mother and what a wonderful person she is. I thought about me with eyes so sleepy I can hardly keep them open, so I got back into bed again to try and sleep.

0615, I am awake again. I cannot sleep any more tonight. I hope tomorrow night will be better.

0630, The least little thing is waking me now. There were real voices outside my door and I came wide awake again, but experiencing no palpitations or anxiety. I just became wide awake.

0635, Asleep again.

I lay on my bed until it was time for my bath and then I relaxed as I soaked in the bath at 0700.

0720, After my bath I did a load of laundry and sat in the lounge reading until 0730.

I became very tired and went to my room and slept until 0810. I am feeling more alert now. I had breakfast. I am finding it difficult to read, so I did some woodworking. I talked with Dr Goughi and she reassured me that there was nothing too much to be concerned about, and that most likely, I would get discharged on Monday after spending the weekend at home, hopefully uneventful. She felt my sleeping in short intervals is due to inactivity and strange surroundings. She wants me to take an Ativan before I retire this evening to see if that will make me sleep better.

I spent the rest of the morning at fitness and crafts. I had a nice lunch. After lunch, I went to relaxation therapy and then varnished my pieces of wood. This is quite different from the Waterford Hospital. I have things to do!

After that Dad and Max visited. My family are so misinformed and mixed up with the diagnosis. They feel they are helping me, but really it is making matters worse.

1700, Dad, Paul and I left the hospital and went to my home. After supper Paul and I picked up his daughter from dance school, and then we went to a movie.

I can sense a difference in Paul. The warmth and closeness I felt with him before is gone. He is very cold towards me now, and I do not like that. I suggested that we not see each other until I get discharged from hospital and feel better; otherwise we will destroy whatever feelings exist between us. Our relationship may not survive this upheaval.

Paul left in agreement with this. I know I am strong enough to stand on my own two feet alone if I have to. I have come through this year alone and I can get through the next one alone.

I took an Ativan on the advice of Dr. Goughi and settled down for a good night's sleep around 2230.

Nurse's Progress Notes
February 26, 1997, 0000-1600

She slept poorly overnight, approaching nursing station several times to let staff know that she was awake.

She became anxious, stated she was having "palpitations"

Ativan 0.5 milligrams sublingual given at 0500 hours with relief. Routinely observed.

0800-1600

Pleasant upon approach. She attended occupational therapy today.
Patient sitting in lounge reading this afternoon for awhile.
She reports feeling improved. She discussed some stressors.
Support was provided. No pressure of speech noted.
There is no evidence of increased activity.
She was seen by Dr. Goughi today.
She had visit from her brother and father this afternoon.
Routinely observed.
She is socializing and interacting with co-patients.

1600-2130

Patient is very upset and tearful today over visit with her father and brother.
She says they think she is bipolar and needs medication.
Patient disagrees with them.
Patient went on a pass with her boyfriend to see a movie.
She is still sad upon return.

Thursday, February 27, 1997

0130, I woke up to go to the bathroom, which I did. I am itchy all over and cold. Now I will go back to sleep.

0135, asleep again.

I can't sleep so I thought I would do some writing.

0215, I am trying to sleep but very itchy.

I can't sleep. It seems that I am worried about everything. I am not having palpitations or anxiety, but I am itching all over.

The nurse brought me in face cloths and towels to wash and put some more ointment on. This is really a first!

I am trying to sleep, but now I am cold, like there are chills all over my body. My head is very clear and my thoughts are of understanding. It seems that my family and Paul are misunderstanding me and they feel the answer to my problems would be drugs. I definitely do not agree with them.

I feel I am a very strong person and the answer for me is to get my life back on track, where it can be worthwhile and rewarding, not struggling like I have been because of being diagnosed with a mental illness and fighting for financial and emotional survival.

I have just made myself a promise. I am not going to try to explain myself or what has happened in the past two years to anyone. The more I tell, the more confusing it must become to others. The least said right now to anyone will benefit me the most.

Yes-I want to be well and strong again-but my future is not as bright and clear as I would like it to be due to being labeled with a mental illness and people not believing in me-not even my own family-except Karen and Brian.

Brian does not come often to see me because he is afraid the doctors will ask him questions about me that will be misunderstood. He does not want to be responsible for any part of any diagnosis, as he knows now what happened when collateral history was incorrectly taken at the Waterford Hospital. He is a very solid emotionally stable person who supports me. Both my children are standing behind me one hundred percent because they are the only two people who have known me and lived with me all their lives, except for the past three or four years.

Their father also believes in me, which is also good for them to have that support. I believe in myself and that is what I am most concerned about right now, my immediate family and me.

I have to get better and nobody but myself can help me do that completely.

I know why I became ill. This set-back was stress induced because I was totally backed up against the wall with no income, no job, and no hopes to get one, due to being labeled with the moniker of mental illness. It is enough to drive anyone crazy.

0140, I could not sleep as my legs and arms, and under my arms are very dry and itchy. I am scratching my shins to pieces. I went to the desk to tell the nurses and she gave me some ointment to rub on them.

0300, I just woke again. My earring that I was wearing got hooked in my blanket, so it took me awhile to free it. I still want to sleep without medication so I will try again.

0700, I missed Dr. Frecker, but Dr. Goughi saw me. I had a long visit with Dr. Goughi who told me she does not believe I am manic depressive and that I would be able to go out tomorrow for a weekend pass, return on Monday, and be released.

I do not require drugs. They cancelled Ativan and all I am taking is Premarin. I believe the Ativan was causing my itching.

I did wood working and crafts, relaxation exercises and drawing therapy. The days here are quite full with activities.

Well today I have been so busy taking care of myself I have had no time to write. I went to bed around midnight. I did not realize it was so late. We were having a conversation about drugs and their effects. Of course I was contributing my experience regarding the many side effects that I have experienced while on certain drugs.

Tomorrow I will be going home after I see the doctors. Hopefully I will finish all my crafts before then. I am really looking forward to being able to sleep in my own bed and get back to normal living.

Nurse's Progress Notes
February 27, 1997, 000-0800

Approached nursing station at 0200 hours complaining of generalized itching, no rash. She couldn't identify anything that would cause this.

She was advised to have a cool sponge bath, which she did and obtained some relief. She is napping thereafter. Routine observation continued.

0800-1600

Noted to be up and eating in the lounge.
She ate well for breakfast.
She approached desk to ask to be seen by her nurse.
She was told she would be seen a little later.
Returned to room for short time and attended occupational therapy. Morning medication was given.
Patient was allowed to express her thoughts about her illness, her battle to get back to the work and her reluctance, if she did return to work due to stress that she might be continuously watched.
She was seen by Dr. Goughi re patient viewing her chart.
Pleasant, continues to complain of some itching on skin. [1]She is feeling it was due to Ativan she had last night-wanted to read about the drug. Routinely observed. Passes ordered.
Out to lounge for supper. She continued to be pleasant upon approach.
She is to go on pass later. Routine observation.
Client returned from pass, enjoyed same. She is socializing in lounge.
Explaining to co-patient the theories on how different medications work. i.e Lithium.
She Settled for sleep, but awake frequently.[1] She complained of itching on arms and legs.

---

[1] Compendium of <u>Pharmaceutical and Specialties</u> (CPS) pg.241 Allergic Skin Reactions, while taking Ativan

Client questioning if her clothes were washed in ivory detergent as she has sensitivity to same. No rash seen.

Client out frequently to desk, suggested sponge bath for cooling effect. Client settled for sleep at 0400.

Friday, February 28, 1997

When I got in bed I fell asleep immediately and woke up about an hour or so later "on fire"- It seemed like my whole body was itching with reddened areas appearing on my legs and arms that were very hot to touch.

I wanted to get into the bath to "cool off" but I was not allowed because of the time of night it was. I stuffed myself limb by limb into the sink and coated my skin with body lotion. I put on a hospital gown got into bed and tried to go back to sleep. I was both shivering and burning at this time, but finally drifted off to sleep

Clinical clerk's Notes
February 28, 1997

Assessed with Dr. Frecker today. Doing well.
She is complaining of middle insomnia.
[1]She refused to take Lorazepam (Ativan) because of increased itchiness for two nights.
Otherwise normal and appropriate behavior.
Attending occupational therapy and enjoys that.
Attention and concentration good. No thought disorder, no pressure of speech, mood not depressed euthymic, affect reactive, no suicidal thought.
She is planning to continue her current courses at Memorial University.
She did well on short day pass on February 27' 97. She is looking forward to her weekend pass and possible discharge on Monday or Tuesday.

Plan: 1) Observation
    2) supportive therapy
    3) weekend pass, unaccompanied
    4) Possible discharge on Monday
    5) follow up with Dr. Doucet March 6,'97

Nurse's Progress Notes
February 28, 1997, 0800-2000

She is pleasant upon approach.
She was seen up and out to lounge for meals. Routinely observed.
Patient signed out for weekend pass. Pass medications given and explained to patient. Pass is unaccompanied. She is to return on Sunday.
I was at home on a pass for a few days.

Nurse's Progress Notes
March 3, 1997, 0800-1600

1230, Patient has not returned from pass as of yet.
She was due back at 1200. Dr. Porter is aware.
1240, Writer called patient at home. Patient reports she has been doing laundry and will return when she has finished same. Dr. Porter is aware.

Patient returned from pass 1330. She reports she had a good weekend. She is asking for discharge. Patient wants pass to go to funeral. Patient left unit at 1400 to attend same.

Dr. Frecker notified re patient wanting discharge and will visit later. Patient again late on her pass. She was due back at 1500. 1540 patient still not returned. Dr. Warren is aware.

1635

Patient returned from afternoon pass. She is anxious for discharge. She is waiting to be seen by Dr. Frecker. Patient admitted to feeling good. She highlighted her weekend pass.

Patient says she has major stressors in her life but is not going to let it bother her for the moment. She says she will be alright because her disability payments will not be cut off. Patient says she will take one issue at a time.

Patient inquired as to obtaining a copy of her chart while a patient here. Patient encouraged to discuss this with Dr. Frecker.

Patient to see Dr. Doucet on March 6.

Dr. Frecker saw patient and discharged her to home.

Patient left unit at 1700 hours, in good spirits. Routinely observed.

There was no evidence of bipolar disorder and the discharge diagnosis was adjustment disorder

How ill the doctor fares, if none fare ill but he. /Philemon

St. Clare's Hospital Psychiatric Unit on the 6th floor

Chapter 16

Finally Back to Work
April, 1997

I came home from hospital March 3, 1995, and spent most of the first few weeks reading, writing and relaxing.

I had been under extreme stress now for over two years, especially the last six months since I have been trying to get back to work. My stress level had been reduced considerably since I was informed during my hospital stay at St. Clare's that my disability benefits would not be discontinued until I was back to work in my job at Health Science Centre, working full time hours.

I felt that the anxiety that landed me in hospital this time was "stress induced", so I researched "Stress" and wrote the following:

<u>Understanding Stress</u>

Stress is a fact of life, whether it is positive or negative stress.
I had a good stress tolerance before this unresolved issue of mental illness.
I thrived on stress and kept in control of my life.
Stress was increased when I perceived entrapment.
Stress was increased when I lost control of my life.
Stress can push a person over the edge.

I have always known how to manage stress, but I could not manage what was causing this stress as other people namely my employer, the physicians and my siblings were controlling me. You may think that nobody can control you unless you let them, but this theory does not apply when you are a "psychiatric" patient. You lose your control over everything.

I thought this was the problem that resulted from being controlled by psychiatry: I received this diagnosis of bipolar disorder, and then came maltreatment with drugs that have made my condition worse. I have gone through many losses; loss of my mother, loss of my credibility, loss of my career, loss of my finances, loss of many family relationships due to lack of understanding by my siblings, and now the loss of my physical health completely by forced injections with large dosages of dangerous psychotropic drugs. Now my body has been damaged. Hopefully it will not be permanent.

My "hypothalamus" told my pituitary gland to tell my "adrenal glands" to make the stress hormone "cortisol" to help cope with this severe and prolonged stress. The more stress I experienced, the more cortisol the adrenal glands made and the first visible sign was my large abdomen. This stress hormone is now in my saliva. If it keeps increasing, my gums will start to bleed.

When one's body gets overwhelmed with cortisol, there is an increased amount in the blood stream. The immune system is enhanced under this stress, but only for so long. If this extreme stress continues, the whole system will eventually break down and the immune cells will not keep up their fight any more. This will lead to more infections, colds, and flu.

My belief is that if one does not take control of one's life and decrease this stress, it may lead to cancer. With such a family history (on both sides) for cancer, I have to balance my lifestyle and control this stress, or else I will not survive. I love living: I want to survive.

Stress is a vicious cycle because if it becomes more severe, and too much cortisol circulates in one's body, it causes brain cells to shrink, especially the "hippocampus", where new memories are encoded and stored to be recalled later. Hopefully it will correct itself before permanent brain damage occurs.

I must get this stress under control, before it controls me! The body has a great deal of resilience and it repairs itself well. I will overcome this and I will get well and stay well.

<u>How will I manage my stress?</u>

I know what I want and will not settle for anything less.
I know who I am and will be content with myself.
I know I can conquer this stressful situation.
I will stop feeling hurt and angry.
I will relax more and worry less.
I will forgive myself and not feel guilty.
I will forgive others as this will help me to recover.
I will love myself and be happy.
I will be positive and eliminate all negativity.
I will not put myself down as there are enough people doing that already.
I will make it none of my business what other people think or say about me.
I will keep being assertive with increased self-esteem.
I will not react to other's thoughtlessness.
I will improve my ability to trust as the world is still safe.
I will not withdraw from others, resulting in others withdrawing from me.
I will be open, warm and honest.
I will be truly myself without fear of rejection.
I will smile at people and smile at myself in the mirror.
I will really lighten up now and laugh as often as I can.
I will promise this will be the way I will live for the rest of my life.

When I had visited Dr. Wayne Button a few days before I was admitted to St. Clare's Hospital, He had arranged a consultation with the psychiatrist Dr. John Doucet, for March 6, 1997, to see if he could possibly help me get back to work.

Dr. Doucet's dictated initial consultation sent to Dr. Button:

Thank you for referring this fifty one year old separated nurse who is presently living in her own home here in St. John's.

She stated the reason for her visit was that she felt she had been unfairly treated and misdiagnosed by previous psychiatrists and desired to "clear her name" and to get back to work.

She states she was diagnosed as having a bipolar disorder. She denies problems with mood swings and feels this is completely inaccurate.

In May 1995 when she was hospitalized at the Waterford Hospital and certified, she described the experience as extremely hurtful, frustrating, and very stressful.

Her mother died during the time that she was in hospital, and this certainly added to her level of stress.

She did have a number of other stressors throughout the year including developing hypothyroidism while on Lithium and then being hospitalized further at the Health Science Centre in November of that same year.

Most recent stressors have been the fact that she has been off work and unable to return to her work. Her disability insurance will be expiring soon and she feels trapped with regard to her financial situation.

She states these stressors have combined to cause her to be significantly anxious and have difficulty with sleep for weeks. She was not treated with any medications.

She has been on no medications since August of 1996. She states that generally has felt well except very stressed at times regarding mainly these matters as named above.

She denies mood swings. Her sleep, appetite and energy have been usually good. Her interests are well maintained. She denies any symptoms of psychosis. She denies any panic attacks.

Past Psychiatric History:

She's had four previous hospitalizations under different psychiatrists. In 1983, she was hospitalized at the Health Science Centre under Dr. Jain with a diagnosis of drug induced psychosis secondary to cold and flu medications.

In May of 1995, she was hospitalized at the Waterford under Dr. Karagianis and was diagnosed with a bipolar disorder manic state. Prior to this admission she was on Prozac and also on cold and flu medications. She was certified during this admission and was eventually started on Lithium. She continued on the Lithium from May of 1995 to July of 1996.

Dr. Jain had prescribed Prozac 10mg. as he attributed my fatigue to depression, when in fact I had always denied feeling depressed.

She had a further hospitalization at Health Science Centre in November of 1995. She was again on Lithium and Prozac prior to that admission. Discharged diagnosis was bipolar disorder. She became hypothyroid which was felt to be secondary to the Lithium and the Lithium was discontinued by Dr. Tong.

She was given a trial of Tegretol, but that resulted in very poor energy and this was discontinued. She has been off all medications since that time. She has seen Dr. Hogan and Dr. Nurse for independent assessments.

She was hospitalized at St. Clare's last week with symptoms of anxiety, insomnia and fatigue. She states she was not placed on any medication.

Past Medical History:

She's been in good health except for her past diagnosis of hypothyroidism secondary to Lithium. She has not required any medication for this.

Medications:

Estrogen

Allergies:

None known

Family History:

She denies a family history of psychiatric problems

Personal History:

She grew up in Sibley's Cove, Trinity Bay, one of six children. Little early personal history was obtained on this interview. She has, however, worked as a nurse at the Health Sciences from 1981 to 1995. She was working as an instructor in the Operating Room. She has two children from her first marriage, a son and a daughter.

She was married again for seven years and has been separated since 1994. She did some courses towards her Bachelor of Nursing at the University of New Brunswick in 1994 and did quite well. She is presently continuing some of her courses here at MUN.

She does not smoke, abuse drugs or alcohol.

Premorbid Personality:

She describes herself as energetic, intense, and enjoys people.

Mental Status Exam:

She was neatly dressed and well groomed with excellent eye contact.
Her speech was perhaps a little pressured but was of normal volume and rate.
She stayed on track well and showed no formal thought disorder.
Her mood she described is neither depressed nor manic.
Her affect is very appropriate. There are no delusions or hallucinations.
There are no cognitive deficits. Her intelligence is at least normal. Her insight is fairly good.
She does not feel she has any current psychiatric problems.

Impression:

This is a very interesting and somewhat complex case. While she has had previous diagnosis of bipolar disorder and has obviously been quite ill on a couple of her hospitalizations, she disagrees with this diagnosis and is angry with previous psychiatrists for 'labeling her" and placing her on medications which have affected her health and her career.

While she may indeed have an underlying bipolar disorder, it is of note that her hospitalizations with symptoms of mania and psychosis have usually been concurrent with her taking Prozac and or cold and flu medications. She may have had a drug induced mood or psychotic disorder on these occasions.

While she does report some periods in life where she has become fatigued with poor energy and requiring increased sleep, she denies past history of depression. It may be that a diagnosis of cyclothymic order should be entertained as well. (Cyclothymic means marked swings in mood, but within normal limits.)

At the present time, she is feeling rather stressed about her financial and work related issues. Her mood seems quite euthymic at present and I hope she may be able to get back to work in the near future.

(Euthymic means normal)

Dr. Doucet was following me once a week and I looked forward to his visits as they were positive and encouraging. I really thought that he was convinced that I was misdiagnosed with bipolar disorder and that he would ensure that it would be changed on my records, but instead, for some reason I could not understand, he went along with the other doctors.

Dr. Doucet's Clinic Notes
March 14, 1997

At times feeling well but stressed easily (Illegible) …. external factors.
Memory impairment fluctuates.
Doing a university course and increasing problems with concentration and memory.
Energy good and good interest in am, but fluctuates. Feels slowed down. She used to be very organized and very energetic.
Main current concern is decreased concentration and memory.
Discussed current stressors- financial.

Feeling trapped by her inability to get back to work and her disability insurance expires this month.

Plan: **No medication**

> See in two weeks

For those two weeks I occupied myself by mostly reading and writing. I also began to study analytical questions and logic, trying to improve my concentration and memory.

I also did my aerobic, weight bearing, and stretching exercises as usual to maintain my strength and energy, while keeping up with my housework and gardening. I also honored my scheduled appointments with Dr. Doucet.

I have been feeling more relaxed since I have been discharged from St. Clare's and having long talks with Dr. Doucet, who is giving me the encouragement that there is some light at the end of the tunnel, and I will be able to return to work soon, without having to take any medication.

I was trying to follow my own advice on how to manage stress, but it was very difficult for me to forget what had transpired for the past two years. It had been over a month since I have been discharged from St. Clare's Hospital, and I felt I was ready to get back to work.

Dr. Doucet's Clinic Notes,
March 27, 1997

> **History of hospitalization with diagnosis of bipolar disorder.**
> **Past and current major stressors.**
> **She has been off work since May 1995, mad and angry at what she feels was a major error in diagnosis and management with severe repercussions for her.**
> **Currently out of work and frustrated by barriers to return.**
> **Also discussed at length her hospitalization May, 1995 - certified and in Waterford when mother died --- very traumatized by that experience and remains angry.**
> **Encouraged to work toward resolution of these conflicts in a productive way that does not increase her stress.**
> **Also to work towards understanding of her symptoms at the time and her present symptoms with a view to maximizing her health and level of functioning.**
> **She is more hopeful of a return to work. She feels a gradual improvement in her stress and improvement in coping.**
> **Concentration and memory is improving though, but not to premorbid levels.**

Plan:

> Support
> **Return to work in two weeks.**

Dr. Doucet wrote a letter to Blue Cross
April 4, 1997

> Ms. Day was assessed by me initially on March 6, 1997 and was seen again briefly March 14, 1997. She was not under my care during her recent hospitalization, but rather was hospitalized at St. Clare's Mercy Hospital under Dr. Frecker.
> Regarding her diagnosis, she has had four previous hospitalizations under different psychiatrists. In 1983 she was hospitalized at the Health Science Centre under Dr. Jain with a diagnosis of drug induced psychosis, secondary to cold and flu medications.

In May 1995 she was hospitalized at the Waterford under Dr. Karagianis and was diagnosed with a bipolar disorder, manic state. Prior to this admission, she was on Prozac and again on cold and flu medications.

In November 1995 she was again hospitalized at the Health Science Centre under Dr. David Craig with a diagnosis of bipolar manic, she was also again on Prozac prior to this admission.

She remained off medication since that time and showed a gradual, but not stable improvement.

Earlier this month she had rather persistent problems with anxiety, palpitations and insomnia and was admitted to St. Clare's Hospital under the care of Dr. Alan Frecker for one week. She was not treated with any medications and her symptoms slowly improved.

I do not have a discharge summary from this hospitalization and the formal diagnosis is not known to me.

Why I was prescribed Prozac again in 1995 by Dr. Karagianis I have no answer. It is not recommended to give both Lithium and Prozac together, but this practice is done regularly.

I spoke with Dr. Doucet about my plan to return to work full time with a week's orientation and he agreed that I could do that. I was really looking forward to it.

Dr. Doucet's Clinic Notes
April 7, 1997

Ease back program discussed with Judi. It would mean only able to work 2 hours a day initially and gradually increase hours and duties.

This was unacceptable to her and felt she should return to work full time hours with a period of orientation.

She feels well enough for this.

A letter regarding return to work full time provided.

I thought it was better for everybody, especially me, that I return to work "full time" as there was no ease back program initiated by the Health Science Centre, and besides, if I were given a week's orientation, working full time hours, I could catch up with the changes in my workplace. I thought I would be able to perform my duties satisfactorily, as I was very familiar with my job description, working in that capacity for over twelve years. I had never considered my work a stressor. I enjoyed going to work and was proud of my accomplishments there.

Dr. Doucet's letter to Dr. Griffin, Occupational Health, April 3, 1997:

Re: Judith Day:

The above named patient was reviewed by me. She has been off work for medical reasons. She is fit to work in her position as Clinical Nurse Educator (Nursing Instructor 11) as of April 7, 1997.

I feel she can return to work on a full time basis April 7, 1997. She has been off work for an extended period and will require an orientation.

Sincerely,
John Doucet MD., FRCP (C)

I photocopied the letter and personally delivered it to the Staff Health Office, the Personnel Office and the Operating Room, then left immediately to go out of town, as I did not want to be home to answer the phone to be told I could not return to work on Monday. I was determined that my mission to get back to work was going to be possible. I thought that once I arrive there, it would be successful. When I returned home after the weekend, there were messages on my message manager from work, but I did not retrieve them.

On Monday morning April 7, 1997, I went to work only to be told by the supervisor in the Operating Room that I would have to leave after two hours because I was on an Ease Back Program.

My supervisor and I had known each other personally for years, and I thought we had respect for each other. But now, she pointed to the chair and ordered me to "sit" like you would a dog and I said politely "I am not a dog and I do not want to be treated that way."

I told her that I wanted to work eight hours with Wendy, the person who had relieved me for the past two years, so I could catch up on everything. That would enable me to take over my responsibilities by the end of the week. If that is not suitable and I have to go home in two hours, they would have to physically remove me from the premises. I was determined that I was not going to leave until I had worked eight hours as I was not being subsidized by my insurance provider as there had been no ease back program arranged.

I felt emotionally drained from fighting for six months to get back to my job that I loved. I thought once I got back, I would gradually regain my credibility and my employer's confidence in me. I wanted so desperately to get on with my life as it was before May of 1995. It would be difficult but I knew I was capable of doing it.

After an interview with the hospital management, I was allowed to stay and work the full eight hours with Wendy. I thought I was doing quite well after being off for such a long period of time. The one thing that I did notice however, was that I was slower than usual and my short term memory and concentration were failing me. Wendy told me she had also noticed the difference in me from when I had worked before. My memory and concentration were excellent until May 15, 1995, that was before being subjected to psychotropic drugs, namely Haldol and Lithium.

During my next consult with Dr. Doucet, he wanted to do some neurological testing, so he ordered an electroencephalogram (EEG) that was performed a week later in the Diagnostic Neurophysiology Department at the Health Science Centre, on April 14, 1997.

EEG Test Result

Provisional Diagnosis:

51 year old woman with ?of organic mood disorder.

Clinical Interpretation:

This EEG record is abnormal. It suggests focal disturbance of function involving the left and right temporal regions. If epileptic seizures are suspected, a repeat recording after sleep deprivation may be helpful.

Dr. Doucet phoned me at home to tell me about the EEG results and it was then he told me that he felt my problem was neurological not psychological. I thanked him for the phone call. Then he asked me what I was going to do now and I answered that I was going out to do some gardening. He said "that's good."

Another EEG (sleep deprived) was ordered and booked for May 22, 1997. The night before, I had to stay awake all night. Paul stayed with me and I would never have been able to stay awake by myself as he had to keep arousing me numerous times, or else I would have dosed off.

I have always had difficulty staying awake all night. In nurse's training, I spent my third year mostly in the operating room where there was no night duty.

In the morning Paul accompanied me to the EEG Department, and I was taken in immediately to have the test done. I kept falling asleep and had to be aroused continuously.

Provisional Diagnosis:

Follow up sleep tracing

Impression:

This EEG continued to show potentially epileptiform abnormalities in both centro-temporal regions, but now predominating in the left side. The sleep pattern showed no change and the tracing is similar to a previous tracing #97-091

I had an EEG done when I was admitted to Health Science Centre in 1983, which was normal. Now after drug therapy, they are abnormal. I have brain damage.

Result of EEG, 1983

Diagnosis;

Investigation of psychosis

Impression:

The patient appeared to be extremely drowsy during the recording. There is no evidence to suggest a lateralized or focal area or cortical damage or epileptiform abnormality.

I believe these changes in my EEG's are due to Lithium and other dangerous psychotropic drugs eg. Haldol 10 mg. which were forced injections of large amounts that I did not require.

I had been back to work for a month now and had not missed any hours, except when I had the EEG's done. I was working under different circumstances due to three hospitals being incorporated and change of management, but I really loved my work, so I adjusted accordingly.

The new Supervisor managing the operating room whose philosophy had to be obviously different than mine was, "trying to stay within a tighter budget, due to cuts and closures." Mine, being in education, was "the patient comes first, and others can wait." But we both had to be realistic and work towards reducing the budget.

To make matters worse for me, I was personally under different management as well. My new boss, in charge of staff development, was a previous psychiatric nurse from the Waterford Hospital. She may have thought that what she was doing was best for me, but she kept treating me like a child, which was causing me excessive stress.

For the first time during my thirty year nursing career, my competence was being doubted and I felt that these people thought I was inept. My boss told me that all correspondence that I prepared had to be reviewed by her before it was distributed. I had to write my activity reports monthly, instead of every three months, which was what I had done for the past twelve years. Monthly reporting was a complete waste of my valuable time as writing reports and compiling statistics are time consuming and I did not feel it was accomplishing anything, other than perhaps a way to keep me under control.

My office had been moved to a glassed-in observation unit in the recovery room, so not only was everything I did monitored by my immediate supervisor, I was on display to everyone else as they walked by. I sensed that I was being watched very closely. I did not want to become paranoid about this, so I went along with the management's decisions and kept working and doing my job and doing it well!

That summer I spent a great deal of time working on my computer revising the procedures to suit the three hospitals that had merged, transferring revised documents by computer to the other clinical educator, who was covering both the Grace Hospital and St. Clare's Hospital, until the Grace closed.

I thought we were working well together and were certainly getting things accomplished effectively, as this computerized method cut down on travel time between the three sites, which certainly reduced the costs of making these revisions.

We were also meeting with the administration of the School of Nursing regarding the curriculum for the peri-operative nursing program. My job was very interesting and I was challenged everyday with new projects and ideas, which kept me motivated and interested.

Because I was so busy and enjoyed what I was doing, I completely forgot about being on display and kept doing my work while embracing the changes. I definitely could meet the challenges as I had so much knowledge, experience, and enthusiasm to do so.

There was a little setback though that I faced each morning. The screened in cubicle next to this monitor room (my new office) was where the psychiatrists came daily to administer to their patients the controversial electro convulsive therapy (ECT).

While observing this procedure, I was constantly aware of the torture that I had experienced in the hands of psychiatry. In other words, I was being seen by a psychiatrist nearly every day as they walked by my glassed in cubicle to administer electroconvulsive therapy, ECT, which I call an electrical lobotomy. It is known to wipe out people's memory like pressing the delete button on your computer.

Despite this daily disturbing occurrence by psychiatry, I kept my composure, did my job, wrote my monthly reports, had my correspondence reviewed by my boss, and felt that I would still win someday as I loved teaching and still wanted to do this job, which I knew I had done so well for twelve years before this tragedy.

Then one day I was called down to my boss's office to be informed that they wanted me to do an audit at the Grace Hospital Delivery Room where they did Caesarean Sections, as the infection rates were increasing.

I had done these audits on other occasions at the Health Science Centre (HSC) when we needed to do spot-checks and discover accidental breaks in techniques, so we could change our procedures. This work was not done to place blame on anybody, it was done to improve our quality of care, which could possibly keep our infection rates lower.

We agreed that I would take with me to the Grace Hospital, four student nurses who had come from other hospitals across Newfoundland for a basic orientation in Operating Room Nursing. I would do my practical teaching at that site. In the meantime it would give me a chance to audit the surgical aseptic techniques at the Grace Hospital for infractions that may result in Patients becoming infected.

I don't know if that audit report was acted upon, but it definitely needed to be. What I saw and recorded were examples of sterile techniques of third world standards. While I was there at the Grace Hospital for that week, I kept informing the nurses about the breaks in technique that I had discovered and showed them ways to correct them. These observations were also good examples for the nurses who I was orientating. They would be able to take my suggestions back with them to their hospitals throughout Newfoundland and Labrador.

In my teachings, I tried not to criticize, but to help and be patient with my learner, as my mother had taught me to do. I was puzzled when I was called down to the office of the Director of Women's Health, Ferial Khan Bursey, because she had received a complaint about me from Dr. Kum, one of the obstetricians, even though he was well aware of my role as quality assurance coordinator as I had spoken to him many times when he was affiliating at the Health Science Centre.

Fortunately Ferial knew me very well from her working as Nursing Supervisor in the Recovery Room at the Health Science Centre. She understood my job as well, as I had also done nursing audits in the Recovery Room. After I explained to her what I was doing there and why, she understood completely. I heard nothing more of that complaint.

There were changes I also had to face in my personal life. Paul and I had never really settled our differences after my discharge from St. Clare's Hospital, so around the middle of May, we mutually agreed to cool our relationship, but try to remain friends. I realized that I was free once more.

After reading the book <u>Always Change a Losing Game</u>, I knew the feeling of loss, which was normal, would subside and I would be fine. Ultimately Paul had formed an impression of me that developed negative feelings and I felt I neither had the time, energy, nor the patience to try to change his opinion of me or rebuild our relationship. It is important that I stay strong and healthy.

One way I could do this was to avoid people who undermined me or who would argue with me about my psychiatric diagnosis, as my family and Paul believed they knew better than I did. I had no energy left to argue or even try to explain. They believed what they were told by the physicians. I knew better. What had been done to me was wrong, and that was that.

I had to let go of things that never will be, because holding unto the impossible is the greatest source of one's pain. Suffering is a choice I can make for myself. I choose now not to suffer. From Dr. Viscott's words "I open my hand and release the world. I am here. I am all I need." (for now).

Around the same time, I was considering accepting the nomination for president-elect for the Newfoundland and Labrador Operating Room Nursing Association (N&Lorna). I had been nominated for the position years earlier, but had declined as I had many family commitments at the time as I was raising two teenagers. Now with my children grown up, and I being single again, felt it was my time to do whatever I wanted.

Since 1983, I had worked with N&Lorna as the Education Liaison for the province, but now I had the spare time to take on more interests and responsibilities. I mentioned this to my manager and she asked me if I were going "manic" again. When I heard that comment, I knew that it was going to be the most difficult task of my life to live down the label of "manic depression" (bipolar disorder.) I understand stigmatization much better now as I have been discriminated against and stigmatized with that label thereafter.

However the nursing staff and supervisor who knew me well were very supportive and friendly. They treated me no differently than before my illness in May, 1995. Many of the physicians were very cold and aloof towards me, however I had a sense that the word had gotten out through the grapevine that I had begun to file complaints against their colleagues to the Newfoundland and Labrador Medical Board. I felt the physicians were freezing me out of my position.

Again, my main reason to write the complaints about my medical care by Dr. Karagianis, Dr. Coovadia, and Dr. Craig was to have the charts reviewed by other psychiatrists so that the errors and omissions that I had found on my charts could possibly be corrected and the diagnosis changed. The last thing on my mind was litigation. I just wanted to get on with my life as I had done before May 15, 1995.

Physicians are a very closely knit profession and supportive of one another, sometimes to the detriment of their patients. I always admired that quality of cohesiveness in the medical profession, and for many years wished that nurses could be more like them. Nurses on the other hand are too critical of each other and even our professional association will protect and support the patients more than the nurses.

The medical and nursing professions go from one extreme to the other regarding patients and their care. There has to be a better compromise so that everyone wins when errors, omissions, accidents, and injuries occur. After all we are all human and we can make an error, but we should not cover it up to the detriment of the patient. When that happens, nobody learns anything, everybody loses and the errors and omissions keep happening.

I have always felt that women in general are not as supportive of each other as they should be. Many succumb to the control of the patriarchic systems in our society. Instead of working together as females to bring forward women's issues in order to have the equality that we as females deserve, we bring each other down, just like crabs do. We have to learn to be more like dolphins, who always support and care for each other, especially in times of need.

I kept working throughout the summer, but I did not spend much time observing or working in the operating rooms, even though I loved my job, and was proud of what I had accomplished there. After six months of exposure to an environment, which was becoming more negative, impersonal and uncaring, it was beginning to affect me physically.

I was having more difficulty with bronchitis and chest congestion, and I was prescribed a steroid inhaler, which I despised having to use.

I was becoming more and more fatigued to the point that when I was not at work, I was in bed. I was not feeling gloom and doom or negative about my future, or low self esteem. I was feeling exhausted. I was experiencing pain in my upper back, numbness and tingling of my extremities, plus stiffness and pain in most of my joints all the time. I had vertigo (spinning around) and losing my balance, tinnitus (ringing in my ears), when I woke up from sleeping and stood upright, especially when I did it too quickly. There were times I had blurred vision. I was going from one specialist to another looking for answers. I was hypothyroid.

I also visited Dr. Doucet in his clinic. He wrote the following.

Clinic Notes, Dr. Doucet
September 04, 1997

Fatigued past few weeks. Tires more easily. Also persistent episodic neurological symptoms. i.e. Numbness in legs. Fleeting, lasting only 10-30 minutes. She has been on Eltroxin (thyroid hormone) past two weeks. She is concerned about possibility of neurological disease and has seen Dr. Ogunyemi and Dr. Maroun. Also has seen Dr. Tong re fatigue and hypothyroid status.

Judi has complained to NL. Medical Board against some previous psychiatrists, and these physicians have responded to the complaints but have asked the medical board not to release information to the patient because it may adversely affect the patient.

Dr. Young has called and asked me to review this information and given my knowledge of this patient, assess whether this should be released to her.

I discussed these issues with her today as well. My concern would be to act for the interest of Ms. Day as her physician and not as a mediator with the medical board.

I will review issues with medical board and Ms. Day again.

On exam today she is complaining of the episodic neurological problems as noted and fatigue. She looks well, bright and pleasant.

She shows no abnormality of speech or thought. Mood is not depressed.

She is coping well and enjoying her work

Plan: See in one month

I made many phone calls to Dr. Young and convinced Dr. Doucet that, after what I have been through this past two years, I know I have the fortitude to read Dr. Karagianis' response to the Medical Board, regarding my complaints against him. Consequently I was finally permitted to receive a copy. I have decided to print the entire letter so that the reader can have the opportunity to understand the entrapment one faces when a diagnosis of mental illness is formulated. With Dr. Karagianis' description of his diagnosis and treatment, I was sure I would be trapped among the severely mentally ill forever.

Dr. Karagianis Letter to Medical Board
April 23, 1997

Ms Day was admitted to the Waterford Hospital with three week history of irritable and elevated mood, increased energy and self esteem, grandiosity and impulsive spending.

On the day prior to admission she experienced a delusional percept and then believed she was responsible for uniting all the nurses of the world. On the day of admission she was escorted out of a meeting after belligerent behavior. Collateral history verified longstanding instability of mood.

On examination at the time of admission she was hostile and aggressive. Her mood was frightened; affect was anxious, grandiose and labile. Her speech was loud. Her thoughts were tangential, circumstantial and incoherent. She kept repeating 'No, no, no" and oh my God for no apparent reason. She had thought insertion and grandiose delusions. She had no insight. She assaulted two patients and was certified. (In my opinion, this was exaggerated evidence)

Ms Day has never accepted the diagnosis of bipolar disorder as long as I have known her, despite my numerous attempts to explain the reasons for the diagnosis and the facts supporting this. (He did not)

One of Ms Day's colleagues told me that Dr. Jain thought Ms. Day suffered from bipolar disorder but did not tell her this so she would comply with her appointments and treatments. (Hearsay)

Ms. Day alleges that I misdiagnosed her. In Psychiatry, diagnosing is more difficult than other branches of medicine since we do not have any tests which can conclusively confirm the diagnosis. All we can go by is clinical information, gained by interviewing the patient and those who know her and by performing a mental status examination, and by observing the patient in hospital.

Often, if a diagnosis is in question, or is being challenged by the patient, a physician will ask another physician to examine the patient in order to obtain a second opinion about the diagnosis and treatment.

In the case of Ms. Day, no less than six physicians gave their opinions about her diagnosis during the acute phase of her illness.

These physicians included consultants. Dr. D Craig, Dr. M Coovadia and myself as well as Dr. D Pratt who was a senior resident in psychiatry with a general medical license. Dr. l. Zielonka, who was another senior resident in psychiatry and DR. E. Elli a general practitioner. All of these physicians agreed with the diagnosis of bipolar disorder, manic phase, so I am confident that the diagnosis is correct.

To add further weight to the diagnosis, I am enclosing a copy of DSM 111-R Diagnostic criteria for Manic Episode. The history obtained from family members supported the A, B, D, and F. criteria. If I recall correctly her daughter and son were with me on at least one occasion when I interviewed Ms. Day in hospital Criteria C was obviously met from observing Ms. Day's behavior in hospital. Criteria E was met as far as we could determine.

Criteria for Manic episode

- A distinct period of abnormally and persistently elevated, expansive, or irritable mood, lasting at least I week or any duration if hospitalization is necessary)

- B. During the period of mood disturbances, (three or more of the following symptoms have persisted (four if the mood is only irritable) and have been present to a significant degree.

    1) inflated self-esteem or grandiosity

    2) decreased need for sleep (e.g., feels rested after only 3 hours of sleep)

    3) More talkative than usual or pressure to keep talking

    4) Flight of ideas or subjective experience that thoughts are racing.

    5) Distractibility i.e., attention too easily drawn to unimportant or irrelevant external stimuli.

    6) Increase in goal-directed activity (either socially, at work or school, or sexually) or psychomotor agitation.

    7) Excessive involvement in pleasurable activities that have a high potential for painful consequences (eg., engaging in unrestrained buying sprees, sexual indiscretions or foolish business investments.

- C. The symptoms do not meet criteria for a Mixed Episode.

D. The mood disturbance is sufficiently severe to cause marked impairment in occupational functioning or in usual social activities or relationships with others, or to necessitate hospitalization to prevent harm to self or others, or there are psychotic features.

E. The symptoms are not due to the direct physiological effects of substance (e.g. a drug of abuse, a medication, or other treatment) or a general medical condition (e.g hyperthyroidism.

NOTE manic-like episodes that are clearly caused by somatic antidepressant treatment (e.g. medication, electroconvulsive therapy, light therapy) should not count towards a diagnosis of Bioplar 1 Disorder.

Schizophrenia could be ruled out by her high level premorbid functioning and the other conditions can be ruled out by the presence of mood symptoms. Investigations did not reveal an organic cause of her symptoms

Ms Day contends that her condition was drug induced. I did a Medline search and found only one report in the literature of mania induced by one of the substances claimed by Ms Day, Phenylephrine. I am enclosing the results of the search. Ms. Day's mania was not drug induced.

My house staff and I spent a great deal of time listening to Ms. Day, as I documented in the chart. Much of the time, Ms. Day was too ill to be aware of the fact that we were listening to her. There were many times that Ms. Day had enough contact with reality to know we were listening, but evidently she has overlooked that fact now, probably she is ill again. (No evidence)

I am suspicious that Ms. Day may again be ill when I last saw her as an outpatient, she had been compliant with her treatment and we were on reasonably good terms. Now she again is refusing treatment, is again projecting blame unto others and has again resorted to excessive letter writing (this time to the board) a behavior which was seen when she was ill in the past.

Ms Day says she did not know she had an illness. Many times I told Ms. Day that she was ill with bipolar disorder and the reasons why, but she refused to accept this, which confirms she has no insight.

My clinical clerk, Iva Smrz unsuccessfully tried to contact Dr. Jain. Dr. Jain could not diagnose Ms. Day from thousands of miles away. Even if Dr. Jain had told me that in his opinion she did not have bipolar disorder when he was treating her, it did not take away from the clinical facts which were undeniably present at the time of my involvement.

I did not obtain a diagnosis from a psychiatric nurse. I obtained collateral history from a former coworker who happened to be a psychiatric nurse who worked with Dr. Jain. I made my own diagnosis. As Ms. Day knows, nurses are not allowed to make a diagnosis, and furthermore a patient cannot reliably self-diagnose.

My house staff and I collected collateral information from Ms. Day's son and daughter, who accompanied her to hospital. I believe that these would be the closest family members to provide useful and reliable information. We also gathered information from her sister-in-law, brother and father.

We specifically did not speak to Ms. Day's ex-husband due to her decision that he would not be reliable, although Ms. Day had the delusional belief that we did speak to him. The people we spoke to were people suggested by Ms. Day. I remember this because she was very concerned about the possibility of obtaining information from unreliable people who did not know her adequately, and at the same time was quite litigious. I received Ms. Day's consent to speak to these family members.

Ms. Day's daughter indicated that Ms. Day spent money she could not afford on things that she did not need. She told us that Ms. Day's co-students in Fredericton noted her to be very active and disorganized thoughts. Ms. Day's son said she had past episodes of hyperactivity and increased goal directed behavior.

Ms. Day's sister-in-law described her as very much goal directed, persistently driven, energetic and impulsive at times. Also she described behaviors in the past which were suggestive of persecutory and other delusions. She said Ms. Day was often very excited and talked fast. In the month prior to admission, she saw Ms. Day twice and noticed increasing energy, and other symptoms escalating.

She said about a year and a half prior to admission Ms. Day was writing a large number of letters to Ottawa and had to be asked to stop. This person was unaware of any previous diagnosis and was describing Ms. Day's symptoms in an unbiased way.

Ms. Day brother described her as an 'impulsive and driven" person. He said she spent a lot of money buying new clothes for her mother, and on one occasion put all the money from her purse into the collection plate in church. On one occasion she phoned him at 3am to implore him to check on her ex-husband. He said she has always been 'on the edge." Four days prior to admission he and his sister said she was again close to the edge.

Ms. Day's father told one of the nurses that Ms. Day was "not well for some time." She always wanted to "take on the world" and is always writing things in order to fix them. He was worried about her coming out of the hospital for a pass.

Ms. Day described her premorbid personality as "driven, hard working and energetic". These inter-episode traits are often seen in bipolar disorder.

With respect to Ms. Day's son who helped us arrive at our diagnosis, he is not a physician and is not qualified to determine when a patient is well enough to be discharged. It was evident to me that four days after admission, Ms. Day's condition had improved considerably while taking her medications, but she soon deteriorated again, after her medications were discontinued (upon her insistence) verifying that she was not ready for discharge.

In regards to Ms. Day's claim of committal with insufficient grounds, she was continuing to display active signs of mania while talking about selling her house at any price. Her judgment at the time was poor as documented in the chart and a serious decision like this should not be taken while in the hospital suffering from mania. We considered that Ms. Day was at risk to the safety of her property. Considering the circumstances, I believe I was fairly liberal in providing Ms. Day with passes to manage those personal matters of which she was capable.

Ms. Day was given medications while hospitalized involuntarily. This is permitted by the Mental Health Act and is done to permit early restoration of a patient's liberty and health. Following discharge, Ms. Day continued to see me for follow up visits and medication.

As an outpatient, Ms. Day took her medications of her own free will. However, because of the emergence of depression and her side effects, her medications were reduced. This was followed by a relapse of manic symptoms requiring admission to the General Hospital.

Ms Day says Lithium caused her to vomit for days. While it is true that Lithium can cause this side effect, a patient who vomits excessively will quickly become Lithium toxic. Ms. Day's Lithium levels did not become toxic and her electrolytes were in the normal range. Ms. Day is exaggerating her claims that she suffered every side effect except death.

At the time of discharge she was euthymic and not complaining of any side effects. Her medications were gradually reduced after discharge, and her complaints of lack of energy and her increased sleep could have been symptoms of her depression returning.

Actually, the inpatient chart documents her complaints of fatigue prior to receiving any medications in hospital. The dose of Lithium was very small and her levels were sub therapeutic, ranging from 0.47 and 0.69. These levels are unlikely to be responsible for the degree of physical symptoms as claimed by Ms. Day. In fact, on August 24, 1995, Ms. Day told me she was losing weight, not gaining.

Fluoxetine, Prozac was started on August 24 because Ms. Day was clearly depressed and she had previously been well on this medication in the past. On September 6, 1995, the next visit, Ms. Day was significantly improved. On that visit she says she brought her friend, Elinor Udell, to explain how the medications were killing her. Ms Udell told me that she saw an improvement in Ms. Day, and this fact is documented on the chart.

On page ten of her letter to the board, Ms. Day states she believes it was the fluoxetine which caused her chemical imbalance, which led to her two admissions in 1995. However, Ms. Day stated that she is not chemically imbalanced. In hospital, Ms. Day stated that her problem was caused by Trinalin. I don't think that it is possible for all of these statements to be true.

Either she does or does not have an imbalance (bipolar disorder). Regardless of the cause, Ms. Day has a mood instability problem and it was reasonable to use the mood stabilizer, Lithium.

In addition, Ms. Day states that she started Fluoxetine without consulting a physician in 1995. If Fluoxetine played any role in her illness the Ms. Day is responsible for the state of mental illness in which she ended up, but is trying to blame my colleagues and I for her misfortune. Indeed, her decision to begin Fluoxetine without a physician's advice is poor judgment.

I am not sure what Ms. Day is trying to allege when she described "torture" at the Waterford Hospital. I wish to assure the board that I provided professional care for Ms. Day, which was more than reasonable and was in keeping with the standards of my peers.

I am sorry that Ms. Day has lost time for work due to her illness, while I was looking after her she did return to work, and this fact is documented in my note of August 24, 1995. I wrote a letter to Blue Cross on November 15, 1996 because Ms. Day was being pressured to return to work early and I felt she was not ready at that time, but her overall prognosis was good for returning to work.

Ms. Day alleges that I have committed malpractice which caused misdiagnosis. In my experience of medicine, the diagnosis comes before the practice. I did not practice medicine on Ms. Day till after she was diagnosed. She was correctly diagnosed after very thorough examinations and collateral history and was given appropriate treatment. I have not committed any malpractice,

At the time of her admission to my service, Ms. Day was extremely ill and in need of urgent treatment to protect her and others. While in hospital she assaulted two other patients, and before admission she not only self-diagnosed depression but also treated herself with antidepressants. Following discharge she took treatment according to my suggestion and her own free will and was not coerced to do this.

I thereby request that this correspondence not be forwarded to Ms. Day. As I have outlined herein, I believe she is ill once again and I believe that reading this material; will likely aggravate her more especially if she is delusional.

Trusting this response will be satisfactory, I remain,
James Karagianis MD FRCP-

During the summer of 1997, I had visited Dr. Wayne Button seven times, complaining of muscle weakness, loss of sensations, upper respiratory infection, chills, bronchitis, laryngitis, persistent low energy, and post infection fatigue as I kept having bouts of flu like illnesses.

In early September, 1997 I went to Toronto to attend an International Operating Room Conference. During my thirty year operating room nursing career, I had attended many provincial and national conferences; this was my first opportunity to attend an international conference. Despite how sick and tired I felt, I wanted to attend more than I can express.

Upon returning from the conference, as soon as I landed home, I dropped my suitcases and went to bed to rest as I was exhausted from attending the conference, visiting my relatives and the plane ride home. I felt like I was getting the flu again.

The last day I was gainfully employed was Friday, September 26, 1997 as I was admitted to hospital on the following Sunday and went back on disability payments after only working not quite six months. After that admission I was medically retired from the Health Science Centre. That last day that I worked, I performed my duties and left my office to come home as usual and was looking forward to the weekend, having no idea what was in store for me.

I had been asked by Aunt Alice's family, who were related to my second ex-husband, if Aunt Alice could stay with me over the weekend while her apartment was being painted. I agreed to that, even though my father was in town and I was over tired. I thought that at least dad would be spending most of his time with his friend Frances in Masonic Park.

Aunt Alice arrived shortly after I returned from work on Friday afternoon, unpacked her things and made herself right at home as she had done so many times before, because I had always made her feel so welcome. We had developed a good friendship over the past twenty years as she stayed with me when she would holiday in St. John's with her husband during the years she lived in Miami, Florida. After her husband died, she moved back to St. John's, until she passed away, October 2, 2005.

Back to my journal

September 26, 27 1997

My father phoned me Friday night to tell me that Uncle Gladstone had been admitted to Palliative care and he wanted me to go with Frances and him for a visit on Saturday as it was my uncle's Birthday. He will be sixty nine years old and now he is dying of cancer.

I told him I could do that, even though I was entertaining Aunt Alice, and I would pick him up at two thirty p.m., which I did and we drove to St. Clare's Hospital. As we were walking in the parking lot of the hospital, I felt Mother Theresa's presence and something was placed over my shoulders. It scared me for that second, but I proceeded into the hospital and to the Palliative Care Unit. I did not mention this unusual happening to them, but proceeded to the chapel and pondered there for a few seconds at the beauty of the scenery in the windows.

We proceeded to Uncle Gladstone's bedroom to discover that Aunt Lillian and Uncle Gladstone were there waiting for their son Kerry and his wife Anne to celebrate his birthday by blowing out one candle, singing Happy Birthday and giving him gifts.

I had brought him a little unwrapped stuffed Care Bear that I had been given at the conference that I had recently attended. I placed it over his left shoulder, near his heart and hugged him, sincerely telling him that everything would be alright. Then I walked to the other side of the room by the door, as I was feeling weak and tired and needed air because it was very warm in the room.

After we were there for awhile, Uncle Gladstone kept staring at me with this look of desperation and smiling like he wanted me to help him get better. I thought of what happened in the parking lot earlier, and I became frightened, so I left the room and walked down the corridor with a few people following me trying to find out what was wrong with me as by this time I was looking pale.

I felt weak and overcome and thought it was best for me to go home. Kerry offered to drive my car home and Anne followed in theirs. As we were driving up Portugal Cove Road, some of my past experiences flashed before my eyes, especially the car accident I had in a tunnel in Montreal. I told Kerry we are going to have an accident. I asked him if he believed in God and he said "Yes I do" and then I said "your father is going to get better."

Immediately, I felt bad about saying that as who am I to say or do such a thing? When we arrived home, I got out of the car and went into the house as I felt very safe there, trying to forget these past few hours.

Aunt Alice was at home with me so we cooked supper and spent the evening together. Ford visited for awhile. After he left, we watched television, talked, listened to music and went to bed early.

Written from notes in my journal documented and dated Tuesday, September 30th, 1997.

Sunday September 28, 1997

We slept on and off that night and around 0500 as I was looking out of the window in the grey bedroom; I saw the most beautiful double rainbow that I have ever seen before in my life. I asked Aunt Alice to come see it and she was amazed at the beauty also. She told me that she had never seen a double rainbow with that magnitude before either.

During my trial and presentation of my evidence, Aunt Alice signed an affidavit that confirmed that she did indeed see that double rainbow with me as it was written on my medical records that I had a visual hallucination, which I did not.

We slept until 0800 and then, after our baths, we went downstairs and made cream of wheat for breakfast. The rest of the day begins to get distorted and I do not remember so clearly. I remember a friend and my second ex-husband phoning me and possibly my brothers called around noon. Aunt Alice was talking to dad on the phone. Then after that all Hell began to break loose for me. I don't know what alarmed my family that I was manic, because I was not. I was having an enjoyable relaxing day with Aunt Alice.

First of all, dad and my cousin Gary arrived. Then Aunt Alice's family came and I helped them pack up her things and she left to go back to her freshly painted apartment.

When I came back downstairs, I saw two police officers with guns strapped to their hips and two male ambulance drivers behind them at my door, everything began to close in on me as I knew by this time they had come to take me away to where?, Waterford Hospital? I panicked like I had never panicked before. I went back upstairs and was talking to my brother Jim on the phone, asking for an explanation, and I did not get any satisfaction from him so I said goodbye, hung up the phone, and went downstairs to face the music.

I remember losing control of myself, crying, screaming, going hysterical and If I had been an aggressive or a violent person, I believe I could have been injured or worst still shot, trying to defend myself as my home was being invaded, and I felt very threatened by this incident.

At that point, I had no idea who called for the ambulance or why. I did not know exactly what had happened for this to be taking place or what to do or what was going to happen to me. But I did know for sure that if they all left and I had one person to stay with me, despite this trauma and torture that had just happened to me, I would have been just fine.

My cousin Gary who is six foot five inches tall wrapped his arms around me and I wished I could stay there lost in his arms forever. I was so scared and once more defeated by all this confusion that I could not comprehend what had happened for the police and ambulance to arrive at my door. Gary convinced the police to please leave and sit in their car, which they did, while he and Paul calmed me down and coaxed me to go to St. Clare's in the ambulance, which I felt I had no other choice but to do.

I was imagining what the neighbors would be thinking of this incident as I live on a quiet cul-de-sac where we all knew each other as we had lived there for years. I felt more defeated than ever now as I am going deeper into this psychiatric trap.

While riding in the ambulance, I was definitely saying and doing things that were not making sense. I was driven crazy as the saying goes.

I reflected on the previous ambulance ride, which happened February of this year, from Old Perlican Hospital, but why were they taking me away from my home in an ambulance now? I am still haunted by these overwhelming experiences of being controlled by others with my self-empowerment stripped completely from me. This frightening incident, that I remember so well, has helped me to understand how misunderstood and misjudged people feel, for example the wrongly charged and convicted, who are incarcerated unjustly and who subsequently lose control of their autonomy and their lives.

It is amazing how people with a psychiatric label, which is placed on them so recklessly by psychiatrists, are then considered incompetent, unreliable and dangerous and need to be controlled.

After I had some time to reflect upon this episode, I believe now that Aunt Alice was exaggerating to my family about me and my behavior, as she was known to have done in many situations with others. While she was talking to my father, he misinterpreted her remarks that led him to believe that I was in a manic phase of bipolar illness and he phoned my cousin Gary, who came with him to my home.

I learned that my brother Jim was phoned and he took the initiative to call 911, the Emergency Department at the Health Science Centre and informed them that I was manic, without even seeing me. Emergency responded by sending the police and ambulance. My brother did exactly what members of a family who have a relative diagnosed with bipolar illness are instructed to do. Again I felt I was being controlled, trapped, and tortured.

A physician is a person who works sixteen hours a day telling others to slow down or they'll get high blood pressure.

St. Clare's Hospital, Palliative Care Unit where my Uncle Gladstone spent his last days.

Chapter 17

Second Admission to St. Clare's Psychiatric Ward
September 28, - October 1, 1997

This is also written from notes that I had documented on Tuesday, September 30th, 1997, approximately thirty six hours after this trauma of being forced to go to the hospital by police escort in an ambulance.

Sunday, September 28, 1997

I was placed in a small room in the Emergency Department and I know I was there sitting and waiting for hours and hours, feeling very stressed, confined, desperate, psychotic, overheated with bright lights, and felt I was being tortured. I was feeling the same way as I felt at the Waterford Hospital two years ago in that darkened room with Dr. Karagianis, when I was certified and I had to take drugs or be injected. This present interrogation is lasting for hours.

Members of my family were taking turns coming and going. I remember Uncle Charles and Betty there. I don't know who called Paul, but he was there also. I definitely did not need him there as we had ended our relationship and he was dealing with the loss of his friend who had recently died.

I was feeling like I was in a well lit, overheated torture chamber and was becoming overwhelmed, so they injected me with medication. They took blood for testing and finally I was transferred to my room, which I do not remember.

Emergency Department

Nursing Continuation Sheet

September 28, 1997

1710 Settled in client room 2. Clinical Clerk notified
1750 Dr. Renouf is assessing patient in client room 2
1800 Consulted to psychiatry, Dr. Duke paged.
1810 Patient refusing to have labs ordered.
1815 Dr. Duke notified
1820 Patient allowed labs to be drawn
1850 Dr. Duke is assessing.
2040 For admission, bed requested.
   Next two entries illegible due to poor photocopy quality.

I imagine that is where they injected me with the potent neurotoxic drug Haldol 5 mgs, as it was ordered to be given immediately. They once again forcefully injected me with the wrong drug against my will. Ativan alone would have been fine, or perhaps just some emotional support and understanding. Again they were treating a diagnosis of bipolar disorder, not stress-induced anxiety.

2103 Appears to be less agitated, sitting beside friend, not (illegible)
2130 Sitting in client room number 2 with friend, awaiting assessment by Dr. Callahan.
2230 Report given. Certification Papers not signed. Dr. Callahan paged times two. Transferred with nursing assistant and friend via wheelchair to 3west.

2235 Certification papers signed by Dr. Callahan.

I was confined in that small room for over five hours after being put through extreme stress by the police and ambulance during the apprehension and transportation to the hospital. That in itself had created severe and prolonged stress and now I felt I had been tortured for another five hours.

Medical Consultation Record
September 28, 1997

Identification:

51 year old separated nurses from St. John's who lives alone in her own house.

Chief complaint:

Brought to emergency room by ambulance that was called by her family following a two day history of increased agitation with psychotic behavior. 'I'm God. I'm dead."

I remember saying "My God" I'm dead. And then saying out of frustration in that small room in emergency, I'm dead, you're dead, we're all dead.

History and Physical:

Patient has a 12 year history of psychiatric illness. On this occasion she was noted to be very agitated for two days. She has not slept well, has been noted to have increased energy, and has been repeating "I'm God" and "I'm dead. You're dead. We're all dead." All day.

It seems she was functioning okay until Friday night, when she visited her uncle in palliative care at St. Clare's Mercy Hospital. She has been repeating "I died for my uncle" a lot during the interview.

Her brother is also reporting that she was planning a big Oktoberfest party last night.

I have planned a party around October 27, almost every year since I can remember to celebrate my birthday and Halloween.

She reports seeing "a large double rainbow at 5am. Last night," but denies any other hallucinations. She repeatedly said that 'I am God... I am Mother Theresa" today. There are short intervals when she does know who she is though. She denies depressed mood. Says she is happy.

She denies suicidal or homicidal ideation. Denies ETOH. ?

She has been violent today as well, biting her friend on the right temple and repeatedly smacking and kicking at him during the interview.

As a result of taking Lithium for over a year, I have a reduced seizure threshold (abnormal EEG's). The symptoms of clapping, biting, etc are seen in temporal lobe epilepsy. I believe being as stressed as I was with this forced confinement and ill treatment all day, caused this violent behavior that has been described above, which was definitely out of character for me.

Past Medical History:

Chart notes head injury 20 years PTA.

This is the first time my head injury, (frontal left lobe) from a motor vehicle accident is recorded on my psychiatric charts. I broke a windshield with my forehead in 1976 and sustained head and neck injuries.

Hypothyroidism.

Past Psychiatric History:

Several admissions since 1980s, most recent to St. Clare's Mercy Hospital in Feb.'97. (discharge diagnosis adjustment disorder.)

Admitted to Dr. Craig Nov. '95 (diagnosed with bipolar disorder mixed state.)

Admitted to Waterford May/95 to Dr. Karagianis (diagnosis bipolar disorder manic episode.)

She was treated with Lithium, Haldol in the past.

Has been refusing to take any medications other than Eltroxin and Premarin for months

Past History:

Documented in old charts.

Family History:

Mom has history of depression, father and brother have problem similar to hers. (This is inaccurate information)

Mental Status Exam:

Well groomed lady, looks younger than stated age, psychomotor agitation, and patient couldn't sit longer than minute at a time and repeatedly hit and kicked at her friend. Speech pressured. Mood "happy", affect labile. Patient is agitated, aggressive and confused.

No suicidal/homicidal ideation. She had hallucination of seeing rainbow this morning. She has grandiose delusions of being God.

On one occasion was orientated to place, but not thereafter. Other tests of cognition were impossible. No insight at all.

Physical Exam:

Patient uncooperative- could not complete

Impression:

Bipolar disorder-manic phase with psychotic features
(secondary to stress and no medications)
Sept 28, 1997 2230

Chart received, case discussed with Dr. and patient seen for examination and completion of certification form as per Mental Health Act. Agree with impression and management. Dr. Callahan M.D.

Mental Health Act 1971

Certification

The following facts and reasons for the above opinion of me the first physician are as follows.

51 year old female with a long history of bipolar affective disorder, has not been on any medication for several months, presented to emergency today with a two day history of manic symptoms including euphoric mood, grandiose ideation and delusions (says she is God).

Insomnia times two days, irritable, aggressive, no insight into her illness. When seen, somewhat calmer as she had responded well to sedative medication given on an emergency basis because of her aggressiveness and agitation, but still no insight as to her need for treatment, hence I feel she has to be certified as she needs hospitalization for her own safety.

The following facts and reasons for the above opinion of the second physician are as follows.

Pressure of speech, grandiose, agitated, no insight delusional, believing she is God, unpredictable at risk of self discharge and self harm.

Nursing Assessment

Medications brought in:

Eltroxin 0.5 milligrams, every day, by mouth
Premarin 0.625 milligrams by mouth, every day

Physical disabilities/pre-existing health Problems:

Hypothyroidism

Previous psychiatric admissions:

Health Science Centre, St. Clare's, and Waterford.

Family:

Mother father brother have history of mental illness (inaccurate and escalating)
Has son and daughter who live away. (Inaccurate)

Interests/Friends/Hobbies:

She has several friends. Involved in work related programs
She works at Health Science Centre as R.N. in operating room.

Risk Factor:

Denies

Drug Abuse:

Denies

Suicidal Ideation:

Denies

Other stressors contributing to illness:

Uncle in Palliative care unit
Recent death of friend.

Sexuality:

Not in relationship at present.

Menstrual cycles:

Menopausal

Circumstances Leading to Admission:

Patient sedated on admission and slept as soon as reached nursing unit. Unable to complete nursing assessment at this time. Will do so at a later time

Psychiatric Nursing notes
September 28, 1997 2311

51 year old female patient admitted from emergency with a diagnosis of manic episode under the services of Dr. Callahan. Settled in room 3251, under constant observation. Certified with one signature. Plus, Plus sedated from emergency.
Due to client's condition unable to perform nursing assessment. No belongings brought up with client, only her purse which is locked up in the drawer at nursing station.

Friend of patient explained to writer that she feels she is not sick and does not need to be on any medication. As soon as you mention the word "med' she becomes plus, plus agitated.

Sleeping in room at present.

Remains on constant supervision with RNA (Registered Nursing Assistant.

Medical Progress Notes
September 28, 1997

Settled a bit, but still looks restless, irritable. Speech pressured. She knows today that she is not God.

Monday, September 29, 1997

I do not remember at what time I awoke today, and finally the drugs began to wear off, but I was very drowsy and refused any more medication.

I was informed that I was certified and could be forced to take them, but I said WHY, if I do not need them now? I don't, as I feel safe here and that is how I want to stay, drug free.

The day was very long and I walked the corridors and watched television and made phone calls. I ate my meals. The day was endless. I retired around 2130 hours

Psychiatric Nursing Notes
September 29, 1997

Remained on constant observation overnight. Slept for most of the time. When awake she had no interaction with staff. Asleep at present.

0800-2000

Mainly isolative to her room. Was on constant care initially in morning but this was changed to close observation. Seen by Drs. Frecker and Weiner

Dr. Weiner ordered Epival and Ativan but patient refused these this morning. Patient feels she will get better on her own.

Religious tone to some conversations. Denies suicidal ideation, and denies thoughts of wanting to harm others.

Drowsy in appearance, stated feeling tired. Agreeable to stay in hospital. She requested to speak with pastoral care.

She refused blood work in morning and Dr. Weiner aware. Close observation maintained.

Medical Progress Notes
September 29, 1997 1445

Judi is much more settled today. She says she knows she's Judi today, but does remember everything that happened yesterday. She was very drowsy this morning, but is much more alert this afternoon.

She asked for a pass to visit her uncle in palliative care, and seemed okay (and agreed) when I told her this wouldn't happen for a few days at least.

She says she has no suicide ideation, and will not try to leave.

She feels she is better today and does not want any more pills.

She looks a bit restless this afternoon, and her speech is pressured. She doesn't appear to have flight of ideas, and her grandiose ideation of being God seems to be gone right now.

Impression; manic episode of bipolar 1

Plan: Continue medications, close observation, no pass.

Nursing Psychiatric Notes
September 29, 1997

1600

Refused 1600 hour Ativan, stating she was feeling too drowsy. Dr. Duke CC aware. No further intervention ordered.

1845

Has been up and around unit. Feels she does not need new medications to get better. States she is doing "fine" and that medications will cause her to be too tired. Wondered about a pass tomorrow to go to a "wake."

Advised to discuss this further in morning. Used telephone. Remains on constant observation

2000-2400

Patient approached writer telling her that she does not require medications and that she is now well. An hour later was speaking on the phone – noted to be talking about God.

Refused HS (night-time) medications stating her thinking is "all cleared up" and that she can get well on her own.

Writer spoke about the importance of maintaining administration of her medication in order to stay well – She continued to refuse her pills.

Settled in bed at 2200 hours. Maintained on close observation.

Slept until 0100 hours, requesting paper and pencil so that she could write out her thoughts about the past couple of days.

She stated that she is upset that her family doesn't believe that she can get well without medication.

Patient slept for 10-15 minute naps occasionally during the night. Maintained on close observation.

Tuesday, September 30, 1997

It is 0110 and I have just awakened after sleeping since 2130 last evening. I feel okay and do not have any symptoms except a dry mouth.

I slept until 0130 when I was aroused by the nurse shining a light in my face. I wrote for about twenty minutes, fell back to sleep and was awakened again around 0230 so I am writing again.

It is now 0300. I just woke again scared and wanted to cry because I have to make the nursing staff and the doctors understand that I have to be given space.

My brother Jim phoned tonight saying I have to take the drugs and it started to upset me so I hung up on him. My family members do not understand what I went through at the Waterford Hospital two years ago and what I have just been through again for the second time in six months. This is very hard on my system and it is making me worse.

When I was admitted in February, I did not need any drugs and I was out of hospital in less than a week. My family has to be educated about this and they must realize that they are doing me more harm than good. I know they mean well but they lack the understanding.

I just woke up again and started to cry, but stopped immediately because I have to be strong. It is very difficult having to isolate my whole family and my special friend because they will not listen to me about "my side" of this story.

They insist the doctors are right and I am manic depressive and that I need drugs. They think I am going against doctor's orders and they are going against me.

I was doing well until this weekend. In fact next weekend I was going to Grand Falls for our Operating Room Nurse's Association meeting as I was running for President Elect. I still want to do this, so please believe me when I say I am well and explain to my family that they are doing more damage to me than good.

I was not allowed to go to crafts or woodworking because I do not have a referral. Do they not realize how this kind of treatment hurts the patients? What is wrong with going for therapy if you feel you want to go and get a referral later. This upset me greatly and I began to cry, but I did not lose control or cry for too long. The main thing now is not to get too upset and lose my cool.

I don't know how long I will be in here, but I am not going to take any drugs as I can't hold my head up when I take them and I am very drowsy. It is not the way I want to live.

I must get out of here by Wednesday as I want to go back to work Thursday and go to Grand Falls on Friday. This admission is unnecessary and wrong.

It is 1400 I just found out that I am certified. That means I am locked up in here for at least two weeks. Dr. Frecker saw me yesterday for a few seconds in the corridor and hardly spoke to me. That is how easily a certification is made. It is incredible!

Julie phoned me from the Canadian Mental Health Association and possibly somebody will visit me from there. I spoke to Dr. Doucet and he cannot come to see me until tomorrow morning. I will survive until then as I know how strong I am. I will be in control and I will be okay.

While I was talking to Julie at the Canadian Mental Health Association, she told me she will help me get out of here. Dr. Doucet will also help me. My family members still believe I need medications and they will not help me get out.

I just phoned Sheila in Corner Brook and asked for her support for my bid as President-Elect for Newfoundland and Labrador Operating Room Nurses' Association. I know I can do the job and a darned good one at that.

At 1605 I am still wondering around the ward trying to keep myself occupied. I played patience for awhile and walked the corridors. The time is going very slowly and I wish I could get out of here. I spoke to Dr. Henley about the matter of being certified.

My cousin Linda phoned me from Toronto and Karen has phoned me twice today. My brother Jim came by to visit me, but I was so upset with him I would not see him.

Psychiatric Nursing Notes
September 30, 1997 0800-1600 1355

Up in morning. Tearful and talkative re events leading to admission. Lacking insight and saying she doesn't need to be in hospital. Easily irritated. Says she feels calm and relaxed inside. Writing on paper throughout day.

She is asking for Dr. Doucet who has treated her for the past six months. Asking for discharge.

Talkative at times and upset easily but says "anybody would be upset if sent to hospital if they were not sick." Asking re: certification

When reinforced to her re certification she became upset and restless. Discussed this with Dr. Weiner and Dr. Duke. Awaiting Dr. Doucet's consult

Patient has been telephoning Dr. Doucet's office and has also been talking to her family.

Reinforced to her the importance of continued hospitalization and medication. She refused Ativan and Epival in morning. Close observation continued.

Patient is aware that she cannot sign herself out of hospital.

Patient is agreeable to stay in hospital until Dr. Doucet sees her either today or tomorrow.

Patients says she is going to stay calm until then.

1600-2000 1835

Used telephone earlier in the evening. She states she is doing "fine" and wants to go home soon. She is eager to see Dr. Doucet on consult. Given reassurance. Had visitors. No tearful episodes noted. Close observation maintained.

2000-2400

Patient is socializing with co-patients. She continues to have poor insight into her illness. Believes she is ready for discharge. Maintained on close observation Settled in bed 2200hours.

Medical Progress Notes
September 30, 1997 0800-1600 1355

Still no insight. Refusing treatment. Says she doesn't need treatment. She says she feels well. Denies delusions and hallucinations. Looks anxious, very (illegible).

As above Patient says she feels well and doesn't need treatment. Refusing medications. She has kept a record of her experiences since Saturday- see copies on chart.

Says she has complaints in to the Medical Board because she has been misdiagnosed as bipolar disorder. She would like second opinion from Dr. Doucet.

The nurse mentioned to me that she was considering signing herself out. I talked to her and reconfirmed that she was certified. When she learned that there were two signatures certifying her, she became very angry and hysterical for around three minutes, but quickly calmed down.

She was very upset, saying she was being tortured again like she was in the Waterford Hospital. I informed her that Dr. Doucet would see her.

Wednesday, October 1, 1997

At 0030 I was awakened again with a flashlight. I went to the desk and pleaded with the nurses not to wake me up so often as I need my sleep because I did not get a good night's sleep last night.

I wrote in my journal until 0045 and went back to sleep.

I feel well rested now and hopefully I will be home tonight in my own bed.

I got my bath at 0715 and got myself ready for the day.

I sat down with TG, one of the patients, for breakfast and we had an interesting discussion about the municipal election.

I saw Dr. Frecker and discussed this past Saturday and Sunday with him.

Psychiatric Nursing Notes
October 1, 1997 0300

Awake for a brief time. Maintained on close observations

Medical Progress notes (clerk)

Saw patient this morning. She was much more calm and controlled, and apologized for her outburst yesterday.

Mental Status Exam:

Casually dressed, well groomed. Good eye contact. Mood subject "calm, feels good." Objectively calm, a bit irritated. She denies hallucinations, delusions. No suicidal/homicidal ideation. Speech a bit pressured, circumstantial. No tangentiality.

Cognition:

Orientated to person, place and time. Recall 3/3. Repeat 7/7 digits. Abstraction good. LTM Good. Attention good (says alphabet backwards). Insight fair.

Psychiatric Nursing Notes
October 1, 1997
800-1600 1335

Up and about unit. Has not settled. No anxiety or agitation noted. Still refusing psychiatric medications. Seen by Dr. Frecker. Patient decertified and granted pass to Health Science Centre to see neurologist.

Patient accompanied there. Neurologist (Dr. Pryse-Philips) to forward letter to Dr. Doucet and Dr. Frecker.

Patient says she feels well and feels her family and friends overreacted on the weekend. Speech pressured at times. Says she doesn't see any reason why she has to be continued on certification.

Says Dr. Pryse-Phillips was going to suggest a medication for her.

Patient is waiting to see Dr. Doucet.

Patient denies any suicide thoughts. She says she is thinking clearly.

She has been reading pamphlet on Epival.

She was encouraged to accept Epival as a treatment option.

Brother visiting at present. Close observation continues.

I went to the Health Science Centre for a consult with Dr. Pryce Phillips, a neurologist who recommended that I should take Epival because of my abnormal EEGs. I agreed to take it upon his advice.

I spent very little time in Dr. Pryce-Philips' office explaining my symptoms and I thought his formal neurological examination was not thorough. In fact he spent very little time with me, and I felt this visit was more of a formality than an actual consult as he also believed that my symptoms were psychiatric.

Dr. Pryce-Philips wrote the following to Dr. A. Ogunyemi, a neurologist to whom Dr. Doucet had referred me for a neurological opinion.

Dr. Pryse-Philips wrote
Re Judith Day
October 1, 1997
Dear Dr. Ogunyemi

Thank you for asking to see this lady. A few days ago her brother came to the house and called an ambulance and took her off to St. Clare's Hospital, where she was admitted in what sounds to be a manic state. I believe she is being discharged from St. Clare's today and she came here with a nurse from the psychiatric ward.

I can find no evidence of a neurological problem in her. Her symptoms being non-specific and variable and her signs non-existent. I really don't think that this is a neurological problem because her mental state I think can explain all the current symptoms that she has- like stiffness of her hands after holding on to things, sometimes depending on position. A feel as though her left leg is dragging sometimes, increased need for sleep, tightness of her interscapular muscles, impaired short term memory and imbalance on getting up quickly after lying down.

The formal examination of the nervous system was completely negative.

The use of Sodium Valporate (Epival) is being considered for her. I would be thoroughly in favor of this because it would help both the abnormal EEG, which is showing spike activity and may have some association with mental state and also undoubtedly her mental state itself.

Psychiatric Nursing Notes
October 1, 1997
1530

After reviewing reading on Epival, patient has agreed to start on same tonight.

1600-2400 2100

Patient using telephone frequently during evening. She is awaiting Dr. Doucet and hoping to be discharged. She states she is feeling good.

No voiced suicidal ideation. Dr. Doucet visited and interviewed patient.

Order written for discharge, Dr. Frecker notified.

Prescription given, patient to make own appointment with Dr. Doucet. Blue card given to patient.

Patient is in good spirits –no delusions noted.

Brother coming to main entrance for patient. Left unit at 2100 hours and discharged home.

The discharge diagnosis was <u>Stress Induced Anxiety</u>.

According to Diagnostic Criteria from DSM -1V-TR, I did not even meet the criteria for Brief Psychotic Disorder as the psychosis did not last for one day. I felt I was being emotionally tortured from the time the police and ambulance drivers arrived at my door around noon until I was transferred to my room at St. Clare's Hospital just before midnight.

DSM-1V-TR Criteria
Brief Psychotic Disorder

A. Presence of one (or more) of the following symptoms.

    1. Delusions

    2. Hallucinations

    3. Disorganized speech (eg. Frequent derailment or incoherence

    4. Grossly disorganized or catatonic behavior.

Note: do not include a symptom if it is a culturally sanctioned response pattern.

B   Duration of an episode of the disturbance is at least 1 day, but less than one month, with eventual full return to premorbid level of functioning

C   The disturbance is not better accounted for by a Mood disorder with psychotic features, schizoaffective disorder or Schizophrenia and is not due to the direct physiological effects of a substance (e.g., a drug of abuse, a medication) or a general medical condition.

Specify if:

With marked Stressors (brief reactive psychosis): If symptoms occur shortly after and apparently in response to events that, singly or together, would be markedly stressful to almost anyone in similar circumstances in the person's culture.

This admission was so traumatizing for me and the circumstances of the forceful apprehension would be markedly stressful, as mentioned above, to almost anyone in similar circumstances in our culture.

DSM-1V-TR, Desk Reference to the Diagnostic Criteria. American Psychiatric Association, pg 161

I had been mistakenly treated so unfairly again, especially when the police and ambulance came to my door and I was escorted out of my home and unnecessarily taken by ambulance to St. Clare's Hospital.

I was confined in that well-lit, overheated, enclosed space of a monitoring room in Emergency Department while being pressured for all those hours, and taking up valuable time of medical personnel observing me. I will never understand this treatment of the "supposed" mentally ill. I felt I was being driven to insanity.

Would doctors go to this great length of such a pressure test to prove that their diagnosis of bipolar disorder is correct? If so, this is wrong and I suffered unnecessarily as a result. Was it any wonder that I became anxious?

I started taking Epival as recommended by Dr. Pryce Philips and I believe this drug nearly cost me my life.

Good medicine always tastes bitter. / Confucius

St. Clare's Hospital, Ambulance Entrance

Chapter 18

Back on Long Term Disability

I came home from the hospital on Wednesday evening, arriving around 2200, October 1st. You can imagine how demoralized I was, but still relieved that I was let out, thanks to Dr. Frecker, who decertified me and Dr. Doucet for discharging me. I don't know what happened to Dr. Callahan or his opinion that led him to place his signature on the certification form, but I was told by my treating physician that there were no obvious signs of bipolar disorder. This information was also written on my chart. If there were signs, I would not have been decertified by Dr. Frecker, as I kept insisting that I had no insight into having the symptoms of bipolar disorder as I don't.

I promised myself that I would take Epival, if I could tolerate it at all, as my EEGs are abnormal, a direct result of the damage done to me by Lithium that I was forced to take for too long.

I had a consult with Dr. Doucet on October 7, 1997. The interesting thing I noticed when I read Dr. Doucet consultation records was that 'Bipolar Disorder' was written on his documents for the first time, after the discharge from St. Clare's Hospital while the hospital records on that admission were reading "stress induced anxiety," or "adjustment disorder." I had been subjected to extreme stress prior to both of those admissions.

Consult Dr. John Doucet
October 7, 1997 1804-1850
Bipolar Disorder

Past two weeks after a flu became fatigued and drained and exhausted, past week again sleeping well but drained and exhausted as day goes on. Problem with concentration. Denies feeling depressed. Looks very tired.

Admitted to St. Clare's Hospital September 25th-not sleeping-symptoms of mania. She is not accepting of this. Feels she was stressed and became hysterical while confronted with family, police and ambulance.

Back to work last spring-after long term disability
Back to work after being off work Sept 26th. Hospitalized from September 28th to October 1st.

Again long discussion of her symptoms past and present and ability to work. <u>Slow</u>? Long term disability

Shortly after the September, 1997 admission, upon the advice of Dr. Doucet and arranged by Dr. Ogunyemi the neurologist, I travelled to London Ontario to see Dr. George P.A. Rice on October 20, 1997.

Dr. Rice spent a considerable amount of time questioning me and I answered his questions honestly as to what I had felt at that time.

Dr. Rice's Consult letter to Dr. Ogunyemi
Re consult October 20, 1997

Thank you for referring this rather complicated patient.
This 52 year old registered nurse was a good historical source and was quite troubled by the symptoms which led to her referral here. She described several problems.

She has been complaining of fleeting numbness, rarely lasting longer than five minutes. Much of this numbness appears to be related to change in position. She denied aggravation of this problem with neck flexion.

I mined the history for remote episodes suggestive of demyelization, (multiple Sclerosis) **but could not find anything convincing. In particular she denied the Lhermitte and Unthoff symptom.**

The Lhermitte sign sometimes called The Barber Chair Phenomenon is an electrical sensation that runs down the back and into the limbs from involvement of the posterior columns and is produced by bending the back forward or backward.

The Unthoff symptom is the optical nerve involvement where eyesight is lost temporarily, especially when the temperature is increased.

Back to Dr. Rice's medical consult:

She denied episodes of vertigo, loss of balance, unusual fatigue or sphincteric disturbances. I had episodes of all those symptoms. I was surprised that Dr. Rice had written that on the report or else my memory had become worse than I had imagined and did not relay this information to Dr. Rice.

She also described a substantial change in memory. She claims that she can no longer perform her job as an operating room nurse. Her memory for recently acquired information is failing her and she has been troubled by inability to complete her daily routine without tremendous effort and exhaustion. Her trouble at work has been noticed by others. She also described increasing difficulty in maintaining passing grades in extracurricular studies.

She also described a peculiar tendency to sudden outbursts. On at least four occasions she has suddenly and without provocation had outbursts during which she would behave in an inappropriate manner. On two occasions she was found to be uttering expletives. On each occasion she has had no memory for the episode once it terminated. Each of those episodes had prompted short term hospitalizations. The only identifiable trigger had been anxiety. She mentioned that she recently had an electroencephalogram (EEG), which revealed some spike wave abnormalities, although I did not see a report directly.

In the last fortnight, she has been started on Epival. She was taking Premarin and replacement levothyroxine.

These behavioral irregularities have been attributed to a manic depressive psychosis, although treatment with Lithium has not helped.

She claims that much of the recent fatigue originated with a severe flu like illness which she experienced in May 1997.

The family history is blameless. She is currently separated from her husband and denies that there is any substantial burden of stress attributed to the separation.

She did allude to some ongoing difficulty with her encounters with the Newfoundland medical profession. Some difficulties appeared to have stemmed from a complicated hospital admission.

Investigation with CT and MRI has failed to illuminate the cause for her multifarious symptoms.

Neurological examination revealed mild impairment in recent memory and remarkably prominent difficulty with calculations.

She has good insight into her current problem. Her general knowledge appeared to be normal. Abstraction was reasonable. She had difficulty dressing herself after the examination. The rest of the examination was unremarkable except for a mute plantar response on the right side.

Comment:

1. This was clearly a difficult and complicated case. I was significantly impressed that she was disturbed by a primary disturbance in cognition and I thought that further neuro psychology testing was important. I thought that this could easily be carried out mostly in Newfoundland. Alzheimer's disease and Pick's disease, (another form of dementia) were considerations.

2. The peculiar explosive outbursts could represent a temporal lobe or limbic discharge. I would be interested to see the EEG reports. An extended trial of Epival, perhaps at a higher dose, could be interesting.

3. I would like to review the MRI directly.

4. Some elements in the history suggested the syndrome of Chronic Fatigue, which can be associated with cognitive dysfunction. I thought that further neuro psychological testing would be important here to help with the distinction.

Dr. Rice had mentioned that the only identifiable trigger during those admissions had been anxiety. He failed to mention the times I was admitted while taking Epival, Prozac and Lithium, Trinalin, Halcion, Gravol and other antihistamines and decongestants. Most of these drugs will negatively affect memory.

After my medical consult was finished I spent some time with my cousin Linda in Brampton and celebrated my fifty second birthday with her and her family. That night, just before I went to bed, I told Linda that Uncle Gladstone is going to die on my birthday as we had called and he was so ill. I was awakened by the phone ringing and I had a feeling that he had died. I did not get up and Linda did not come to my room, but the next morning his death was confirmed. Uncle Gladstone died on my birthday.

I arrived back home exhausted from the plane ride and it took me days to recover from the trip. I had a consult with Dr. Doucet on November 4, 1997.

Consult Dr. John Doucet
November 4, 1997
Bipolar Disorder

Fatigue increased this past several weeks. Exercise tolerance poor. Difficulty getting out of bed. Has had the desire, and interest though maintained.
Sleeps twelve hours, which is excessive for her.
She has been on Eltroxin 0.05 mg times 10 weeks.
She has been having worsening physical fatigue.
She has increased cognitive problems with concentration and memory.
She continues to read though and follows well.
Had consultation in London, Ontario, with neurologist – referral Dr. Ogunyemi.
Judi does not accept the diagnosis of bipolar disorder though they may be causing all her symptoms.
Affect and behavior today are appropriate.

Plan:

Get TSH Thyroid Stimulating Hormone and other routine blood work done.
At this point my muscles were becoming more and more stiff, painful and weak. I literally had to roll out of bed. My neck flexion was becoming very limited; eventually I had to start physiotherapy as I was warned by a neuro surgeon I could possibly lose my drivers' license, if I could not turn my head any more than I had demonstrated to him. I went to physiotherapy for six months and since then I have been doing neck exercises to maintain my flexibility.

The episodes of vertigo were increasing to the point when I got up from lying down, I would have to be very careful that I would not fall over as I would lose my balance. I had fallen many times since I had become so ill.

I could not walk on uneven surfaces, like on the side of a hill as I would feel as if I were falling. I could not walk in a straight line and found it difficult walking with a friend, as I would keep walking into her as we walked side by side. My friend would complain laughingly about that, but I could not help it. Difficulty to tandem or walk side by side was a symptom of cfs/fm/mcs that I had found out about later on, once I discovered what was wrong with me and educated myself about the illness.

The way I walked, (gait) changed to accommodate my lack of balance. I was walking very slowly, with my feet further apart and shuffling. I was not lifting my feet or using my heels and toes properly, and my steps were shortened. I was wearing laced up boots as my ankles were weakened and I was constantly tripping over my feet and twisting my ankles.

I was dragging my left foot and climbing stairs with great difficulty as the up and down strength in my legs was declining. I would have to pull myself up by grasping the handrails, even pulling some of them off the walls.

My daughter was the person who actually brought my failing memory to my attention. She told me she would phone me one night, have a conversation and call me again and I would not remember that she had even called or what we talked about and I would repeat things over and over. The Epival that I started on September 24, 1997 was making my symptoms worse.

The Lhermitte sign or electrical sensation is also seen as part of "discontinuation syndrome" associated after taking certain psychotropic medications such as selective reuptake inhibitors, particularly Paroxetine and venlafaxine for long periods of time that are suddenly discontinued.

The Unthoff symptom that Dr. Rice referred to in his report is the optical nerve involvement where eyesight is lost temporarily, especially when the temperature is increased.

All I can say about my vision is that it has always been changing day by day or hour by hour. My mother took me to an ophthalmologist when I was a child because I was complaining of blurred vision. At that time, she was told that I had inflammation behind my eyes, whatever that meant, but I did not need glasses.

I really had blurred vision though, when I was forced to take Haldol and Lithium, during the final few days that I was at the Waterford Hospital May-June 1995; even though it was documented on my medical records I had no complaints upon discharge. That last day I was locked up, I said nothing about my physical symptoms as I just wanted to get out of there and deal with them later.

Since I have been a child I have always complained that "my eyes made me feel sick", especially when things are moving around me and there are strange spots and prints in view.

I kept my appointments with Dr. Doucet as I was requested to do by Blue Cross.

Consult Dr. John Doucet
November 20, 1997
Bipolar Disorder

> **Past week or so not as extreme in fatigue. She is coping better with it.**
> **She is exercising more regular. Mood good.**
> **Discussed at great length current stressors.**
> **Feeling labeled with psychiatric disorder, unable to work, etc.**
> **Discussed anger and frustration with this stressor and feeling "hurt" By past difficult experiences.**
> **She looks well. A little over talkative possibly but mood stable.**
> **She is determined to focus on her wellness.**

As much as I was focusing on getting better, I felt I was getting worse. My muscles were becoming more painful and stiffened and I was getting weaker and weaker and I was retaining more fluid.

When I took Diazide for the fluid retention, I would feel very ill, and get headaches, so I took only little pieces of the drug and only when I absolutely needed to, as I could not tolerate them.

It was around that time that I had an appointment to see Dr. Wayne Button who passed on to me the diagnosis of fibromyalgia and gave me a hand-out with information on that illness. Even though I was a nurse, I had never heard the word before; therefore I was not familiar at all with it. I was puzzled, but at least I had a diagnosis other than bipolar disorder, but I still kept my appointment to see Dr. Doucet.

Dr. John Doucet's Consult
December 16, 1997

Fatigued. Trying to exercise regularly but fatigues easily with exercise and walking over stairs. Muscular pain and weakness.

Sleep at times excessive.

Weak and fatigued. Muscles often painful and stiff.

Diet good, using vitamin supplements.

Interests fairly stable. Has had increased problem with fluid retention.

She is on Diazide (diuretic) once a day, but takes it once a week.

Eltroxin 0.05 mg.

Mood and interests are very good. She denies ever having any depressive symptoms. Accepts that she has severe stressors past and present.

Fatigue predominant. She fears getting ill.

Discussed at length, past and current conflicts and stressors.

Looks fatigued at times, but optimistic about her capacity for coping and for future wellness.

Karen wanted me to come to Fredericton and spend Christmas and New Years with her. I thought it would be a good idea, even though I did not look forward to the flight. Flying had become very difficult for me as I would become exhausted and it took me a long time to get back to normal afterwards, whatever normal had become for me.

Karen and I had a really wonderful time that holiday season. We shopped and bought each other red and green satin pajamas that we wore Christmas Day. Karen did not have any Christmas decorations at that time, so we decorated ourselves instead with our red and green pajamas and lounged around all Christmas Day reflecting on other Christmases while watching the fireplace video on television, and unwrapping our many other gifts that we had for each other.

Since I was now diagnosed with Fibromyalgia/Chronic Fatigue syndrome and knew very little about the illness at that time, except for the handout Dr. Button had given me, I bought a book <u>Fibromyalgia and chronic Myofascial Pain</u>. A Survival Manual by Dr. Devin Starlanyl and Mary Ellen Copeland and I began to read about this newly diagnosed illness.

After reading the first few chapters, I realized that I have had this illness for a long time, and I knew then that I had been misdiagnosed, and the drugs that I have been taking for the past two years have made my condition worse. By then, I also understood why I had been so sensitive to the drugs that I have been given to treat bipolar disorder, and I have to somehow find a way to get myself better, or else I will become totally disabled.

On New Years Eve, Karen and I went out to a party, but I had to come home early because when many of the patrons lit up cigars, I had a reaction to the cigar smoke. By the time I picked up my coat, called a cab, and got out of the building, I was violently ill.

Another evening Karen suggested that I get into a relaxing bath, with aroma therapy, which she had made for me and I became so weak that I had difficulty getting out of the water and into my bed.

By now I was also noticing, I was having difficulty swallowing, and food would get stuck in my esophagus. I would always have to have a glass of water handy when I ate. Occasionally I would vomit up a large amount of undigested food.

I was not feeling well for days after that exposure to cigar smoke, but my flight to return home was booked for January 5th, after Karen's birthday. Despite feeling ill, I wanted to go home as Karen had to go back to work, and I did not want her to have to worry about me. I have always looked after myself and I thought when I was back home I would feel better.

I thought once I got on the plane, flew to Halifax and from there fly directly to St. John's; I would be home before I knew it. That was not the case. There was a delay due to bad weather in Toronto, and the plane did not arrive in Fredericton. There was no definite time announced to fly, so feeling sick and tired, I just stayed at the airport and waited and waited for many hours for the plane. I did not bother Karen or even let her know I was waiting as I knew she was very busy at work and she would worry.

Finally we boarded and flew to Halifax to change planes to St. John's. After spending more time in Halifax, I did not get home in St. John's until nearly midnight. It had taken seventeen hours to get home. By that time, I had gone past the point of exhaustion and I could not sleep however hard I tried.

I was being compliant with taking Epival as I wanted to ensure that I would not place myself in jeopardy of having a seizure, but however hard I tried to relax and sleep, I could not. After another two days and nights of restlessness, I went to the Emergency Department again as I was afraid I would go psychotic if I did not get any sleep. I had absolutely no racing thoughts. I was exhausted. This is clearly not the way a person who is bipolar-manic would present herself to emergency.

Health Science Centre
Emergency Department Record
Jan 7, 1998 0810

Notice the difference below in the two assessments done by nursing and physician

Nursing:

Complaining of being weak, shaky, generally unwell for past few days. Concerned it is depression or her neuralgia. Vitals 120/74 P. 78 R.18 T.36

Physician:

Patient of Dr. Doucet complaining of extreme anxiety.
Unable to sleep past two or three days.
Her mind is racing.

When I presented myself to emergency on January 7, 1998, I was assessed at the Health Science Centre, the impression was bipolar hypomanic. Dr. Doucet visited me and allowed me to go home and rest after giving me a prescription for Haldol 2mgm. Haldol was not what I needed at that time as I was not hypomanic I was overtired and unable to sleep.

Dr. Patel in Emergency Department wrote:

52 year female with bipolar disorder, extensive history, abnormal eegs. On Epival since two months ago. Complaining +++ruination. Re diagnosis of fibromyalgia made two weeks ago and her implications of her illness, why diagnosis was not made earlier.

She is able to put those thoughts away, but they rapidly return with +++ anxiety. She has slept poorly this past three days.

She has apparently psychotic episodes in the past and is afraid of her present situation may precipitate another.

Except for poor sleep, denies manic.

Except for decreased energy, denies depression. No suicidal ideation. Self care. Admits to anxiety.

History:

Multiple previous admissions. Diagnosed with bipolar disorder. Has had psychotic episodes. Tends to be noncompliant with medications and does not accept diagnosis of bipolar disorder.

Past medical history:

fibromyalgia, hypothyroidism, 2 abnormal EEGs, suggestive of epileptiform activity, endometriosis.

MSE

Appropriately groomed and dressed. Eye contact. Speech pressured.

+++ tangential, preoccupied with fibromyalgia, psychomotor agitation. Affect labile. No hallucinations or delusions. No suicidal ideation. AGO times 3 Recall 1/3 at 3 minutes.

Insight and judgment fair.

Seen by Dr. Doucet. She was discharged home with a prescription for Haldol 2 mgs p.o qhs (at bedtime).

Advised to rest. Will call Dr. Doucet in am

IMP: bipolar, hypomanic.

I went home and tried to rest and relax in order to feel less tired.

I was still taking the Epival as prescribed, but I believe that this drug was bringing me down to the point of severe exhaustion. I was so tired, I was wired. I did not need more medication. I needed understanding and support which I was not receiving from anybody.

Back to my journal:

January 7, 1998

I came home from the hospital and relaxed all day. I went to bed to try to sleep without taking Haldol as prescribed. I was still taking Epival at night so I thought I would rest and go to sleep, which I did and had a pretty good night's sleep.

January 8, 1998

The next night, I wanted to do the same, but I could not fall asleep, therefore after trying for awhile, I took the Haldol. I awoke in the middle of the night panicked because of shortness of breath and palpitations. I went to Emergency at Health Science Centre around 0630 on January 9, 1997.

Health Science Centre
Emergency Department
Nursing:

Patient took Haloperidol (Haldol) last night,
She has not eaten for 3 days.

> **Fibromyalgia. Under care of Dr. Doucet.**
> **She woke up from sleeping with palpitations, Shortness of breath.**
> **Patient anxious and upset. 0650 138/88 p 88 r 28 t 36.7**
> **1115 112/70 p 86 r 20 t 36.7**
> **52 year lady, psychiatric patient**
> **Severe anxiety attack. Saying she is going to "die"**
> **Requests to see Dr. Doucet. Consult psychiatry**
>
> **0735, Patient chanting I'm dead, I'm dead. Clapping her hands, rubbing her legs and feet.**
> **She can't eat solids. The food is not digesting.**
> **0740, With some reassurance she took Ativan.**
> **0815, Taking fluids.**
> **1115, Resting quietly. In no apparent distress.**

I was not even ordered to have an EKG done, despite having palpitations upon awakening, after I had taken Haldol with Epival. This combination of drugs had caused me to wake up with palpitations and shortness of breath,

According to the Compendium of Pharmaceutical and Specialties CPS,

[1]Precautions re Haldol: Because of the potential to cause orthostatic hypotension through alpha adrenergic blockade, haloperidol should be used with caution with patients with cardiovascular disease.

I have had ST abnormality on EKG's since the admission to Health Science Centre in 1995 and have not been ordered another EKG since the summer of 1996. I have suffered from orthostatic hypotension, (blood pressure dropping from changing to upright position from lying down) for quite some time now.

Haldol may also lower the seizure threshold and should be used with caution in patients with a seizure disorder or a history of drug induced seizures, abnormal EEGs or head trauma. I had all the above. Why was this drug Haldol ever ordered for me to take at that time? Why was I so stupid to take it? I have no answers. However "I believe now, my brain, due to neurotoxicity, had been more damaged than I want to know."

For months in 1995, at the Waterford Hospital and again at the Health Science Centre I had been prescribed Haldol with Lithium, in doses that I could not tolerate, which had caused damage to my brain.

Now, according to[2] CPS, "reversible cerebral atrophy and dementia have been reported with Valproate (Epival) therapy."

Dr. Doucet told me that I needed to be hospitalized and I took his advice. I wanted to be admitted to the neurological ward at the Health Science Centre as my chief complaint was palpitations upon awaking, extreme fatigue and muscle weakness and pain and now I was not digesting food.

He then advised me that that the psychiatric unit was the best place to be admitted as they would have the staff to observe me. I agreed to be admitted to psychiatry, believing that these doctors knew by then that my diagnosis was fm/cfs/mcs. Unfortunately, they had not been educated about this condition and I was still being treated as a psychiatric patient.

I wonder why ye can always read a doctor's bill, an ye niver can read his purscription./ Dunne.

---

[1] Compendium of Pharmaceutical and Specialties, CPS, 2008 pg. 1027-1028
[2] Compendium of Pharmaceutical and Specialties, CPS, 2008 pg. 817-819

Christmas with Karen, 1997, gifts galore!

Chapter 19

Admission to St. Clare's Psychiatric Ward
January 9-12, 1998

Friday, January 9, 1998

I was transferred to St. Clare's Psychiatry Unit from the Health Science Emergency Department. I was interviewed by the nurse, but I was too tired to answer questions properly, therefore I came to my room and began to sleep.

Dr. Singh interviewed me while I was lying in bed half- asleep. I am still very sleepy. I slept for two solid hours and awakened very refreshed. I drank soup and yogurt for supper as I still do not want any solid food. It is impossible for me to digest it. Food stays in my stomach for days and then I vomit it back up undigested. It is as if my digestive system is slowed down to almost a complete stop.

I had tea for a snack then took a bath and settled for sleep at 2300 and went to sleep immediately and slept soundly.

For the next two brief admissions I have decided not to document all the repetitive information from the previous admissions. I will focus on the relevant information pertaining to my present symptoms.

St. Clare's Hospital
Medical Consultation Record
January 9, 1998
Focused on various muscle aches
Complaining of palpitations and shortness of breath on awakening.
Slow moving.

Neurology Decreased power throughout – symmetrical (question effort) (According to CPS, Epival is known to increase musculoskeletal weakness. but I am still considered a psychiatric patient.)
Plantars decreased, decreased
Sensation grossly normal

I was so weak at this point that I had to roll over and roll out of bed because my core muscles were not strong enough to allow me to even attempt to sit up normally. I could not breathe properly if there were too many heavy blankets on top of me. The doctors were questioning the effort I was putting into strength testing. My voiced complaints and symptoms were not taken seriously, not even on a positive neurological exam.

Nursing Assessment (Exerpts)
January 9, 1998 1500

Voluntary Patient.
Appearance: Neatly dressed female wearing earrings and lipstick. In bright red blouse and sweater. Plaid pants are too small at the waist.

Physical disabilities/ Pre-existing Health Problems:

1. Fibromyalgia, diagnosed last month back of head/neck/shoulders/ between shoulder blades/ legs. This often causes stiffness according to the patient.

2. Beginning to get Athletes Foot.

3. weakness and numbness in legs

Mental Status: Orientated to three spheres. Eye contact good. Cooperative during admission. Occasionally rubbing legs.

Speech is normal in flow, rate and volume. Tired affect.
Reports short term memory loss.
Denies long term memory being affected.
Denies hallucinations. Reports occasionally gets "weird" when her fibromyalgia "acts up"
Vague about same, but reports "strange' thoughts about God.
Tone grandiose at times.
Began teaching me about fibromyalgia and discussed how many people have misdiagnosed her due to lack of knowledge.
Will now be a "human guinea pig" to help others.

Risk for falls: Unsteady on feet.
Known postural/hypostatic hypotension or dizziness

Medical Progress Notes
January 9, 1998

Ms. Day was interviewed this afternoon and her chart from Health Science was reviewed. She denies having any major problems except poor sleep and stiffness due to fibromyalgia.

She denies any feelings of depression, hopelessness or suicidal ideation and does not feel she has bipolar disorder.

She says she went to Health Science Centre because of palpitations.

She has long history, question, depression and bipolar disorder, but she does not want to accept this.

She was treated by Dr. Doucet for the past one-two years but denies being on psychotropic medications.

Mental status Exam:

Lying in her bed appears drowsy. Speech – increased slightly but normal volume.
Thought – circumstantial, but no evidence of active delusion.

Perception:

No history of active hallucinations, but describes having images when she sees herself on the floor with feelings of dying.
Mood subjectively tired, objectively anxious.
No suicidal or homicidal ideation.

Cognition: appears grossly intact, but not formally tested.

Impression:

Question hypomanic/ manic episode of bipolar disorder. (more collateral information needed)

Plan:
1. to get all the discharge summaries from Waterford Hospital and Health Science Centre.
2. Continue Loxapine and Lorazepam as advised by Dr. Doucet but hold Epival.
3. Observe further.

Psychiatric Nursing Notes
January 9, 1998 0800-2000

Admitted to 3 west as a transfer from HSC. Checked for medications and sharps. Both locked up. Vital signs stable. Settled in room and placed on close observation. Seen by Dr. Singh before supper and is asleep at present.

2000-2400 2200

Patient has been pleasant. Affect bright. No signs of depression / talkative and helpful with patients. Appears energetic and bright but no management problems
Answers questions goal-directed.
However refuses 2200 Loxapine saying she doesn't need it and if she does she will come for it.
Walking around unit with pajamas shorts on.

Saturday, January 10, 1998.

0015, I woke up as I was freezing cold. I had no palpitations, but I was shivering all over. I obtained a few more blankets, covered up with help from the nurse and could not sleep so I decided to write down today's activities before I forget them. I am yawning uncontrollably. Maybe I will be able to settle down again to sleep.

0200, I slept for while. I was very cold in the bed and because I was shivering, I was in pain. I went to the desk to ask for some blankets. I was told there were some blankets at the end of the corridor, so I walked down and obtained two of them. At least the nurse came in and helped me by putting the extra blankets on me.

The only problem now was the extra weight of the blankets restricted me from breathing and moving from side to side. I felt I was in a tangle, but finally I went back to sleep somehow on my back, just to be disturbed by flashlights. I had no palpitations when I awoke so that was good.

0400, I asked to take a hot bath as I was very cold and shivery. It was then the nurse came in and turned on the heat. I did not realize that the thermostat was on 'off' or I could have done that myself. Earlier I had put my jacket over the vent to stop the cold air from coming out. I went back to sleep until 0530 when again I was awakened by flashlight. I still have no palpitations or anxiety though.

I am not saying anything to the nurses about their unfriendly behavior because that got me in trouble at the Waterford –good eggs-bad eggs- good nurses-bad nurses-. I can't understand why a nurse would want to be a nurse if she is not empathetic and caring about human beings. It must be a horrible job for them with that attitude.

Nurses have a terrible way of monitoring a patient's sleep by shining flashlights in their eyes. They are awakened and the nurses believe they are not sleeping. Drugged patients may not awaken, but others would.

0715, After writing in Karen's book that I only write in after I am well rested as my handwriting is better then, I went for my bath and stretched my limbs to get the stiffness out. This was relaxing until a patient opened the door on me. I was surprised, but remained calm. As I dressed, I noticed that the band aid left a reddened mark on my skin. I hope I am not getting an allergy to latex. I am sensitive now to every little thing.

I chatted with some of the patients while sitting around the table. M. took my writing pad as she wanted to write me a poem. It is 1445 and she just brought back my pad with this poem written.

> Where interpretations bleed new answers,
> And the rock is the diamond of my soul;
> The livid is my resting place.
> Break not the peoples of your adventure,
>
> Instead, wind your way through the centre of the earth,
> And make free the flower of your bursting.
> Cry not at the cherishing of agony,
> For it will lead you home.
> And to an enlightened Path of Joy.
>   Of Joy
> Get well Yesterday, Judi,
> Love, M.

I ate two digestive cookies and ice cream just before lunchtime. This is the first solid food that I have eaten for days. For lunch I ate grapes, soup and tea.

I wrote a poem in M's little book. This was a poem that Mom had taught me before I left home at the age of sixteen.

> Good name in man or woman, aye my Lord,
> Is the immanent jewel of thy soul.
> He who steals from me my purse, steals nothing.
> Tis trash, twas mine, now his.
> But he who steals from me my good name,
> Steals not that enriches him, but makes me poor indeed.

I have a good long term memory, but l I cannot remember what I had for lunch unless I write it down.

Supper was a bowl of soup and milk. My brother Jim visited and had supper with me. M. played the violin and piano and she also sang. I walked, chatted, read and spent a quiet evening before bedtime.

I took a nice warm bath around 2100 and went to bed around 2200. I slept immediately but was awakened by the nurse's flashlight at 2230. I spoke to the nurse about being completely asleep, but the flashlight wakes me. I will go back to sleep.

I awoke with a dream, but now I cannot remember it, but I had palpitations again. I got out of bed to get the buzzer that was over on the wall to call the nurse. She came, helped me in bed, put up the side rail and left. I didn't get a chance to tell her about the palpitation, so I lay in bed relaxing myself, waiting for her to come back. I began to feel better, so I am sitting on the side of my bed writing this experience because I do not want to forget it in the morning.

2330, I am ready to go back to sleep, but I can't sleep because my legs and feet are burning and itchy, so I spent ten minutes rubbing my legs and massaging them with body lotion.

My mind went back to that poor patient in the Waterford Hospital who came into my room scratching his legs until they bled and nobody would help him. I got reprimanded for coming to his aid. I wonder where he is today.

2340, My hand is getting tired writing, so I had better go to sleep.

1150, As I lay back in bed, I wondered why the call bell is over on the wall and my side rails are up. I hope I don't have to call a nurse in a hurry. Back to sleep.

2400, I jumped right out of my sleep when the nurse walked in on me. I believe I scared her. I got a fright myself, but my heart did not start racing. I turned on the light and wrote the incident in my journal. Back to sleep.

Psychiatric Nursing Notes
January 10, 1998 0600

Maintained on close observation. Patient slept fair overnight, was up several times complaining of feeling cold encouraged to get more blankets.

Patient asked at 0400 if she could get a bath. Patient told it was too early. Staff member checked the temperature and increased same.

Patient feeling warmer, and more comfortable.

No management problems. No further complaints. Closely observed.

0800-2100

Socializing around the unit in red, satin, lounging pajamas.

Refusing Loxapine as she feels she does not need it. Pleasant.

Socializing with several patients, talking extensively. Exchanging addresses with patient in 60B, as well writing out well known poems for her.

Began crying at one point as one of the poems referred to the loss of one's "good name"

She feels this is what happened to her.

However refused to talk about this saying, 'I'm alright, nothing's wrong."

Composed herself after a few minutes, then began talking and laughing with co patients.

2100-2400

Up and about unit. Very talkative. Did settle early to bed, but hypersensitive to noise. Waking easily when checked. (on close observation) she refused night sedation. Insists she does not need it.

Sunday, January 11, 1998

0250, I have been awake since 0120. My legs and feet are burning and itchy. I wanted to get into the bath, but I am not allowed. I rubbed lotion on them, which helped a little.

0525, I am having a bad night. I am itching. My whole body is itching. I have no ointment left to apply. I went to the desk in bad shape. They are going to think I am crazy so I have to be careful. I know that I am screaming, but I am in so much agony and I need a relaxing hot bath.

I went back to my room and I started having pains in my chest. I thought I was having a heart attack, but they went away. I voided lots after that.

I asked the nurse to cover me as I was so cold. She did.

Then I settled for awhile but every little creak or sound was affecting me. I hurt all over, even my jaws are aching.

I called help! help! to the nurses and one came. I asked if she could listen to me for awhile and she said she was busy, I got up then to write down my feelings so that somebody can know what I am going through tonight.

0530, I hope I can settle to sleep. 0645, I slept soundly for an hour with the light on. (nurses might say I didn't sleep because the light was on) I was thirsty so I walked to the kitchen and drank two glasses of water. My left thigh is very painful and tender to touch.

I am in a complete mess. At least I am not psychotic or manic. I hope I can stand it until 0700 to get into the tub and exercise a little in the water to get rid of some of the pain. I voided again. I got in the tub at 0645. I broke the rules, but I am sure that running water is not as loud as me moaning and groaning.

I got in the bath, ran the water and accidently voided, so I had to empty the tub and begin again. I soaked and soaked, exercised and massaged as I relaxed in the bath with my eyes closed.

All of a sudden I got scared out of my skin and screamed because the rubber mat that I had placed on the towel rack slipped off and fell down on me in the bathtub. It really scared me as I had my eyes closed.

I survived that scare so I had to relax all over again. After being in the bath almost an hour, I dressed, came back to bed and slept.

0820, I am awake now and I feel so drained. I am going to get something to eat now and begin my day. I dressed and combed my hair, put my makeup on and ate breakfast in my room.

I phoned Aunt Alice and I also phoned the girls in the basement to water my flowers and take care of my fish. They reminded me that they did not have their rent paid, (I had been away since before Christmas.) so I asked them to please deposit the rent money in my account, which they did.

While I was taking my bath, I stretched, exercised and massaged my legs, especially my thighs. While I was massaging my thighs, it was as if I were bursting bubbles under my skin. This is the first time that I had noticed that sensation.

My both hips are painful tonight; therefore it is uncomfortable lying on my sides. I sound like a real neurotic crock, but even when I put my knees together, it hurts. It is as if I need a pillow to pad between my knees when I lie on my sides.

I am in a very relaxed state otherwise, but when I am almost asleep, the least little noise really makes me jump and I make a noise. I have no control over that. It seems to mainly happen while I am in a half sleep and half awake state. I have had no palpitations tonight though.

<u>Psychiatric Nursing Notes</u>
January 11, 1998

> **Slept in naps. Would cry out. Complaining of stiffness and discomfort. Very dramatic.**
> **Next not legible due to poor quality photocopy.**
> **Continues to state adamantly that she does not have bipolar disorder.**
> **She reports no emesis since admission.**
> **Seems to appreciate frequent snacks as she is trying to increase food intake.**
> **Writes very detailed notes of her day which includes when she (illegible), what she eats etc.**
> **She says she does this to jog her memory as has significant short term memory loss.**
> **Continues to refuse prn (whenever necessary) medications and her Loxapine.**
> **She feels it will cause her "to go psychotic." (Feels Haldol did this previously).**
> **She is not admitting to delusion/hallucinations. Had visitors.**
> **Wearing red clothes as well as red nail and toe polish as it is a "strength colour".**
> **Pleasant. She is socializing with visitors and patients. She had a visitor after supper.**
> **1900,** (Next nursing note/notes illegible due to poor quality) photocopying).

Monday, January 12, 1998

0145, I slept very soundly. I don't know what time I fell asleep, but I was awakened by a toilet flushing. I sleep soundly, but I awake easily.

I am stiff again and it is painful to turn over in bed. I would love to get a bath now, but I have to wait until 0700, according to rules. Everything hurts because the bed is really hard. At home I have sponge on my bed that makes it more comfortable. Tomorrow I am going to be home.

I bet the nurses will chart that I have not slept all night because I have left my light on. It is better tonight as I am not being as disturbed with the flash lights.

0330, the nurses opened the door to check me. I was sound asleep, and I awoke again abruptly. I haven't had any palpitations tonight. I explained to the nurse that these jolts are really taking its toll on me. I told her about the muscle pain as I am so tender now.

0700, I just woke up with palpitations and I have to get in the bath. I felt fine after I slept, but when Dr. Singh visited and started talking to me about mood swings and, bipolar disorder, I did not relate to him. He told me nurses told him I had mood swings because I cried a few times.

There is something drastically wrong with the system when physicians read nurses notes (which are very inaccurate) and formulate their incorrect ideas without even getting to know the patient themselves.

There has to be a better system in place to monitor patients, for instance I believe it is wrong to wake patients up to see if they are asleep or to give them a sleeping pill. The first night I was here the nurses said I did not sleep. I corrected them with my notes.

By not allowing exceptions to the rules when a patient with fibromyalgia needs water therapy for stiff and painful muscles add to the pain. How can I sleep, when each time I try to turn over in bed I am in complete agony, which is only relieved by relaxing in a warm bath before retiring to bed. Sleeping in a strange place is not good at the best of times.

After speaking to Dr. Singh, I was afraid of the possibility of being certified again and being forced on Haldol, Lithium or Epival or something else, which I thought I would not physically survive. I discharged myself against medical advice before that could possibly happen.

I believed that my present state of health was brought on by three months of taking Epival, which was now discontinued by Dr. Doucet and I refused Loxapine and other psychotropic, neuroleptic agents that were prescribed. My body had to once more adjust to a drug withdrawal and to manage on its own steam, which is a difficult transition and it takes time.

Psychiatric Nursing Notes
January 12, 1998

(Next nursing note/notes illegible due to poor quality photocopying,)

0800-1600

She had tub bath this morning. To dining room for breakfast. Sitting in lounge after breakfast.
Patient refused her Loxapine but took her other medications.

Because of my better judgment and previous experience with those psychotropic drugs, I kept refusing Loxapine, and because I was not certified, they could not force me to take it.

According to CPS, Loxapine is another antipsychotic drug that has more precautions listed than Epival, the drug that I believe caused my symptoms that brought me here in the first place and was discontinued by Dr. Doucet upon admission. Some of the side effects of Loxapine are orthostatic hypotension, which I already have, and Parkinson-like symptoms which I certainly don't want. It also causes muscle spasms of neck and face, which is a part of my life now from head to toe. Loxapine is supposed to be used with extreme caution in patients with seizure disorder, and I already have that problem from taking Lithium and Haldol. (1)Loxapine may cause tachycardia, fast heart rate and low blood pressure, which brought me to emergency in the first place.

Psychiatric Nurses' Notes
January 12, 1998 Cont'd.

Dr. Singh was notified concerning refusal of Loxapine. Patient seen by Dr. Singh. (See Medical progress notes for details.)

At 1120 patient approached nursing station and stated she wanted to self discharge. Dr. Singh was in nursing station at the time and aware.

Patient denies any thoughts of suicide or self harm in any way. She stated she wanted to live. She refused to stay in hospital. She stated she is "afraid" she "will be certified." Reassurance provided. Patient denies being sick. Denies being bipolar. Stated the "only thing wrong" with her is fibromyalgia and "fibrofog".

She stated we don't have to worry about her, she is a "survivor."

She is still refusing to stay. Patient reports she will follow up with Dr. Doucet.

Dr. Singh contacted Dr. Doucet's office.

Patient discharged against medical advice. Nurse in charge and Nurse Manager advised.

Medical Progress Notes, (Dr. Singh)
January 12, 1998 1020

Judi was assessed this morning. Her mood appears very depressed and labile. She started crying very easily when talking about her mother.

She herself does not feel that she has any major problems and denied any concern about her own health. She is refusing all medications.

Mental Status Exam:

Casually dressed, cooperative in the beginning but became angry and left the room when I tried to bring her attention towards her labile (unpredictable) mood.

She became verbally loud and refused to talk any further. Speech normal in rate and volume, but at times loud.

<u>Thought:</u> No evidence of active delusional thinking or perceptual disorder:

<u>Mood:</u> Subjectively- denies having any problem with her mood.
    Objectively –her mood is very labile, but mostly depressed and frequent crying episode.
    No evidence of suicidal thought or behavior

<u>Cognitive function:</u> Appears grossly intact though not formally tested because of her uncooperation.

<u>Impression:</u> Question bipolar mood disorder-mixed phase.

<u>Plan:</u>
1. Support psychotherapy
2. To keep trying to convince patient about need to take medications.
3. continue close observation
4. Not certifiable under Mental Health Act at present.

1120

Judi approached me and told me that she wants to sign herself out against medical advice. She was advised to stay in hospital but she is refusing to talk any further.

Currently she denies any feelings of depression or suicidal ideation.

There is no evidence of acute psychotic episode.

She also never had "psychotic episode" since she came into this hospital.

She does not appear any danger to others or property at present time.

Her cognitive function appears grossly intact.

Though her mood may deteriorate further in future she is currently not certifiable under Mental Health Act.

Plan:

She can sign herself out against medical advice.
Follow up with Dr. Doucet. She agreed to see him as soon as possible for follow-up.

Monday, January 12, 1998

When I arrived home from the hospital, I dropped my bags and got ready to go down to the stadium and watch the noon hour skating. Several people came over during intermission and spoke to me. It was very nice feeling for me and I felt good sitting there watching the freedom of the skaters and the waves of cold air on my face as they whisked by.

At this point I did not have the physical energy or the muscular strength to skate, even though the people there were so kind and considerate of me when I did. The more experienced skaters would skate with me when I did not have good balance to skate around by myself.

I had done a lot of skating at the stadium for years now and got to know many of the skaters there who were so friendly and helpful. By skating with another person, I had better balance and had less fear of falling. I found the exercise of skating easier than walking as I did not have to lift my legs as much and the surface was always even. I could just flow around the ice listening to the wonderful music. I enjoyed that very much and spent most noon hours there.

At this time, I was experiencing many unusual physical symptoms, but it was all attributed to my mental illness of bipolar disorder. I understand psychiatrists and psychiatric nurses are busy, but people's lives and their futures are at stake, when they are misdiagnosed and treated inappropriately.

I believe psychiatrists are forgetting the fact that they are also medical doctors. They should not lose the sight of the possibility of other medical conditions and factors that are causing symptoms of anxiety, hypomania, mania and psychosis, like side effects of drugs, which are well documented.

I believe that if psychiatrists follow this narrow pathway and do not attempt to separate mental illness from organic brain disorders or drug induced brain disorders and treat all psychiatric symptoms (mania is a symptom, not always a disease,) with more dangerous psychotropic drugs, without considering other substantiating factors, then their rate of misdiagnosis, which is now 25-30%, is going to escalate and people's lives will be ruined from the effects of dangerous psychotropic drugs that may lead to their untimely deaths..

The Citadel was a book written by a doctor in the early nineteen hundreds that spoke of a patient going mad with an acute mania attack, from thyroid deficiency and was going to be put away in an asylum, but was instead medically diagnosed with Myxoedema and recovered in a few months with thyroid hormone replacement therapy. This patient was saved from the stigma of mental illness and being put away in an asylum, where he probably would have been for the rest of his life. The doctor wrote in this book "In my opinion Hughes is only sick in mind because he was sick in body and was suffering from thyroid deficiency."

Dr. Nigel Rusted, who was a prominent physician in Newfoundland, when interviewed by The Telegram October 28, 2001 had recalled an incident and said "patients get so toxic that they'd become mentally ill." When this happened he would change from surgeon to peace intervener. He went on to say "I had one chap who had a long Tom Razor and he was going to slit everybody's throat in the room. I faced him and I had my fist clenched, ready to nail him if necessary. But I talked to him and he laid it down."

Drug induced psychosis happens and medical personnel are very well aware of this. Why were the psychiatrists so convinced that my problem was bipolar disorder not drug induced?

I had felt physically ill for some time, but now my symptoms of muscle pain and fatigue, muscle weakness, shortness of breath, palpitations were getting worse. Sometimes I felt I was slowly dying when I was trying so hard to live.

Monday, January 12, 1998:

When I came home from the stadium I got busy taking the lights off the trees outside because it is a mild day and I wanted to get it done before the weather turned cold and snowy.

Dr. Young, President of the Newfoundland and Labrador Medical Board, phoned to say he was delivering my documents by taxi. I had given him my personal records for him to read, so he could have an idea of what I am like in character. I also sent him the notes that I had written while I was in hospital this past weekend.

I asked Dr. Young if he would please obtain the hospital records from St. Clare's and compare them. I believe drastic changes have to be made within psychiatry, or patients are going to continue to suffer terribly.

Imagine saying to an "adult" patient "it is terrible you ran water in a bathtub at 0500 and I am going to tell your doctor." How humiliating, I needed a hot bath at that time to help me get rid of the severe pain I was going through, but there was no understanding given to me.

From what I have observed and witnessed; and the patients I have cared for and helped, once Psychiatric Patients are labeled, they are not taken seriously again and can die in an emergency department as I could have possibly done, from the side effects of psychotropic drugs and may never have their death thoroughly investigated either.

Back to my journal:

January 12, 1998

I have been drinking only fluids since I came home as I am not that hungry. Tomorrow I will, with the help of Margaret and Vanessa:

1. *Straighten out my closets and rearrange my clothes by color so I can remember where things are and it won't be so tiring for me to go through the hangers to find something to wear. (I found it tiring and painful to raise my arms.)*

2. *Put away my tree lights if they are dry.*

3. *Take down my Christmas tree.*

4. *Mail Canada Pension Plan application.*

Tuesday, January 13, 1998

0655, I just woke up after an excellent night's sleep, even without ear plugs. When I awoke, I was thinking about the videos that I had watched last night. While I watched my parent's 50th anniversary video, I really noticed how slow my mother was going over the stairs at the Bungalow in Bowring Park. How slow she was to get out of her chair when she spoke. How off balance she was on the dance floor, her left hand appeared abnormal and swollen. I also observed her difficulty while turning around and her inability to walk on uneven surfaces. (She was only in her seventies then and was on no other medication but Diazide for fluid retention and of course insulin).

On January 13, 1998, I kept my appointment with Dr. John Doucet, who I felt always had listened to me, and he was instrumental on my being investigated by neurology, and was responsible for my EEG testing that showed abnormalities. He had told me in the beginning that he felt my problems were neurological not psychological, but later denied he had ever said that.

Dr. Doucet's Consult
January 13, 1998
Bipolar Disorder

Admitted by me to St. Clare's Friday. Discharged against medical advice yesterday.
Concerns about worsening of somatic symptoms.
She is a little pressured and hyper, but pleasant. Some difficulty staying on track.
Mood "fine"— "wonderful"
Affect – Bright disappropriately so.
No obvious delusions.
Fears she may have a connective tissue disorder of Fibromyalgia, due to the pain and muscle weakness.
She does not believe she has a bipolar disorder, However, she did present to ER twice last week because she knew she was not well, and restless, hyper and hysterical—See notes from that assessment.
Spoke at length about her symptoms and diagnosis.
She does agree to take Epival 250 mg. HS- not 250 mg. twice a day as recommended.
I will monitor and see in four weeks.

After visiting Dr. Doucet, I had an uneventful day and went to bed in the restful pink room, but tonight I took the Epival that Dr. Doucet had prescribed for me again as I really wanted to follow his advice.

Back to my journal,

Tuesday, January 13, 1998

I will watch other videos until I get sleepy. Well I watched the video of Sarah's wedding and got a few laughs. When I was sleepy I turned off the video.

Wednesday January 14, 1998

0340, I am awake puzzled. My hips are aching and I cannot turn as I am stiff. I am trying to wait until later for my bath. Sometime after 0340 in the early morning I woke up with pains between my shoulders. I was very short of breath and was having palpitations. I thought I was dying! I reached for the phone to call somebody to help me, and suddenly realized I was not in my own bedroom and there was no phone at my reach to call anybody.

Then I really panicked and struggling as I rolled out of bed, crawled to the other room and phoned my friend Eleanor, to come and help me. I must have sounded desperate on the other end of the line, so Eleanor said she could not come but she would phone my brother. I asked her not to do that and then I phoned my best friend Blanche and Ed.

In the meantime, (I learned later) that Eleanor did phone my brother Jim, who by this time, was very annoyed with me for not following the doctor's advice and take all those medications they wanted me to, but I knew my body could not tolerate them. I was very neuro toxic due to taking Epival for three months. Epival was discontinued while I was admitted to St. Clare's a few days ago, but I had been ordered by Dr. Doucet to restart Epival again last night, which I did. This was a big mistake!

Jim, Blanche and Ed arrived at my home around the same time and I had to go to the emergency again or Jim was going to call the ambulance and of course the police would have to come. Mentioning the word "ambulance" to me was not the thing to do. All I could think of was September 23, 1997, when the police and ambulance arrived at my door.

I felt now that I had no other choice but to go with him to the emergency Department at St. Clare's as my body would not be able to tolerate the stress of that frightening situation of the police and ambulance again. My friends, Blanche and Ed, also came with me for comfort and support.

It is far better to cure at the beginning than at the end. / Persius

My parent's 50th Wedding Anniversary Celebration

Chapter 20

Readmission to St. Clare's Psychiatric Ward
January 14-19, 1998

While waiting in the emergency department at St. Clare's Hospital to be seen, I was having extreme pain in my upper back, so I was lying down across a few chairs. I was weak, in pain, my legs were like rubber and I had the overall feeling of exhaustion. When Blanche put her hand on my back, I screamed in pain. The doctors did not pay any attention to me or try to understand my symptoms, as they did not speak to me about anything or ask me any questions. I was being treated like a dumb uncommunicative animal again.

I was not interviewed by Dr. Ladha, the psychiatrist on call. He spoke to me very roughly, pointed his finger at me, and threatened me with certification to the Waterford Hospital, if I did not get treated with drugs. These words were the same as adding gasoline to a fire and I became really upset and anxious, fearing for my life.

I have become very distressed over my physical health, which has deteriorated considerably over the past few years; especially the last six months and now I have become seriously ill since I started Epival three months ago. I believe that I am heading for a life of physical and mental disability, if I even survive. My poor state of health is being brought on by nothing more than stress and the way I have been treated by the doctors in this province since I returned here from New Brunswick in 1995.

I went hysterical for a few seconds-minute and then was calmed down by my friend Blanche. She and I have been friends since we were two years old. We know each other very well as we have been through so much together.

Blanche gave me this card that I have framed and kept on my walls for over a quarter of a century.

To my Real Friend

**Regardless of
whom I meet
or what I do
or what I have become
It is the friends I grew up with
that I feel
closest to
and that I have the most in common with.
Though we don't see
each other often
when we do
it is as though
we were always together
so comfortable
so natural
so honest**

> I guess old friends
> who know where we come from.
> Who know our backgrounds
> who know our families
> have an understanding
> of us
> that no new friend
> can ever have
> Susan Polis Schutz.

I was escorted in a wheelchair to my room by security, sedated heavily and slept for a long period. When I awoke I felt fine, but I had problems sleeping the next night because of pain and stiffness. Taking baths relieves this discomfort, but I was not allowed to do that.

History and Physical (Excerpts only)
January 14, 1998

> Patient drowsy not able to cooperate with physical exam.
> Impression: Physically (illegible)

0310

Emergency Department Record.

> Had bath 2230, went to bed, woke up cold and having pain all over.
> Tried to contact several people.
> Patient talking nonsense, flight of ideas, agitated and illegible manner.
> Defensive and states doesn't want to be certified.
> Brought in by brother. Patient self discharged Monday.
> History. Manic Depression. Patient talking incoherently/nonsense.
> Phoning family/ friends constantly today and 'babbling'
> Patient agitated plus, plus during interview. Tearful at times.
> Flight of ideas. Discharged because she "did not want to be certified like last time."
> Consult to psychology. Now I have become a "babbling idiot"

0315-0330

> Seen and assessed by Dr. McIssac. Committed to psychiatry.

The Certification under Mental Health Act, 1971. was signed, but with one signature.

Reasons:

Ms. Day has been in Emergency Department three times in last seven days with one short admission. She self discharged on January 12, 1998 after three days stay.

She has been calling relatives in early hours of the morning. This morning she called her cousin, brothers and friends.

She called people at two or three in the morning first.

She has been crying for no reason.

She has been disturbed for at least the past week.

Her conduct is unsafe to herself and maybe to others. Dr. Ladha.

Consultation Dr. MacIssac

"Pain between her shoulders." Anxious

This lady was brought in by her family tonight after she awoke from sleep very scared and anxious and phoned her family and friends to come and help her.

She doesn't really know what she was upset about, but she thinks it may have been the pain she has between her shoulder blades.

Previously diagnosed with bipolar and seen by Dr. Doucet yesterday and placed on Epival every night and was in hospital over last weekend and left against medical advice. She was admitted for a mixed state/hypomania.

She recently has told her brothers that she is an angel sent down from heaven and now the doctors are trying to poison her.

My brother must have been very frustrated with me to give the doctors that information. I remember joking with him that our mother always called me her "Darling Angel from Heaven" as I was always there to help her. She even wrote that fact in her poem "To my Daughter". Yes, I felt I was being poisoned by drugs and it turned out I was. Due to my chemical sensitivities I am like the canary they send down the mine shaft to check if the fumes are safe for the miners.

My brother's comment was added to the chart to give evidence that I was delusional to substantiate the diagnosis of bipolar disorder. I was not delusional: I was physically ill from the effects of Epival, possibly having an impending heart attack, but my symptoms were ignored as I was considered a psychiatric patient.

Impression: Bipolar disorder, manic mixed state

Plan: admit certified.

Nursing Assessment Form:

Involuntary otherwise the same information that was written just a few days before. I was not interviewed as I was too sleepy. This is certainly not the symptoms that a patient would present with who is indeed bipolar-manic.

Nursing Psychiatric History
January 14, 1998

Chief complaint: manic

52 year old manic depressive. Patient of Dr. Doucet, self discharged Monday. Presently manic with agitation, flight of ideas, babbling. Brought in by brother for assessment.

Psychiatric Notes Nursing
January 14, 1998 0700

Fifty two year female admitted from emergency. Family brought her to Emergency room after she telephone them being very upset and anxious. Settled in room 326.

Diagnosis of bipolar mixed state, escorted to 3 west by security, family and RNA.

Psychiatric Notes Nursing
January 14, 1998 0800-1600.

Patient quiet and sleeping. At times making noises and mumbling in sleep. Assessed by Dr. Snelgrove.

Constant care and morning meds held. Patient too drowsy.

1400 patient briefly became alert to sit up and take medications and then lay back. No food taken today.

Psychiatric Notes Nursing
January 14, 1998 1600-2000

1800, Still sleeping. Breathing easy. May hold 1600 Loxapine unless patient wakes. Constant care.

1815, Patient woke and was able to sit on bed. Drank glass of juice and medication given.

Psychiatric Notes Nursing
January 14, 1998 2000-2345

Slept in her bed all evening. Awake at 2230 hours-up to washroom and took some fluids. Quiet but pleasant. Says she feels okay.

Still believes her diagnosis is incorrect.

Still awake at 2340 and took medications with no problem.

Settled for sleep this time. Constant observation.

Thursday, January 15, 1998

0645, I am lying in bed thinking about the last twenty four hours and how physically ill I am and the pain I was suffering that night I was admitted, and I want to write about it, but I am too sleepy.

I am napping on and off. I went to the kitchen for a glass of water. My legs are very rubbery; I can hardly stand on them.

I can't read anything as I am too sleepy. I had a nap for ten minutes maximum, then I woke with my head clearer. I am trying to read again, but I am not having much luck as I have no concentration.

1415, Boring day, feeling sleepy, trying to read the newspaper and I took another Loxapine, very sleepy.

2200, went to bed.

Psychiatric Nurse's Note
January 15, 1998 0630.

Poor night spent. Patient taking short naps overnight. While awake, patient moaning and wanting to get up. Taking fluids well. Constant care maintained.

Ate full breakfast in lounge. Then took bath, felt better after. Dressed in red sweater and slacks. Complaining of feeling over sedated, but pleasant and cooperative.

Assessed by Dr. Frecker and second signature for certification not signed today. Patient was not felt to be certifiable at present.

Patient still delusional as regards to her illness and does not feel she needs these medications and that she is manic depressive.

Feels she is suffering from fibromyalgia and that this condition is giving her fibre-fog which affects her head.

Judi talked about how she was treated in Emergency Department on admission and how certain comments made her feel.

She voiced that she can't remember any of yesterday, but now she is taking notes.

She questioned if Dr. Doucet will visit her and said she had a horrible experience at the Waterford Hospital.

Behavior manageable at present.

Dr. Kennedy aware medications not taken this morning.
Will further interview if necessary. Constant care.

1600-2000

Continue to feel her medications are not the right ones for her condition.
She is reading and talking to writer about fibromyalgia. Eating well.
No management problems as yet.
She did take Loxapine, but not willing to take Epival.
She is certified with one signature. Mood unchanged. Constant care.
Spent the evening up and about unit. Pleasant and sociable selectively.
She took bedtime medications with some persuasion.
Refused to accept diagnosis of bipolar disorder.
Talks only of Fibromyalgia. Took tub bath.
She feels she should have been admitted to neurology unit but says she was admitted to psychology because she was told nursing care is better. Much reassurance and support offered. Constant observation.

Resident's Note
January 15, 1998

Patient seen today by myself and Dr. Frecker. Today very cooperative during conversation. Pleasant but very easily irritated when reference made to Epival use to control mood swings.
No behavioral problems since admission. She has refused Epival on several occasions. She is strongly stating she is well and wanting to go home. Slept well overnight.
Attention and concentration remains poor.
Patient coping with this by taking numerous notes of activities.
During interview describes mood as "good" affect appears euthymic.
No evidence of thought disturbance; Patient denies any perceptual disturbances.
Patient denies any suicidal/homicidal thoughts.
Insight remains poor-does not agree with bipolar diagnosis.

Plan:
Patient wanting discharge but will agree to weekend pass home.
Patient states she will get someone to stay with her over the weekend.
Discontinue constant observation
Weekend Pass to return Sunday.

Friday, January 16

0600 I woke up when the patient in the next bed had her treatment for asthma. I did not get startled or have palpitations. I am okay.
Dr. Frecker allowed me to go home for the weekend and he said I could be discharged on Monday if everything goes okay during that time.

Psychiatric Nurse's Notes
January 16, 1998 0816-1310

Patient seen by Dr. Frecker and Dr. Kennedy. Constant supervision discontinued. Patient continues to deny she is bipolar.
(1) Refused to take 1000 dose of Loxapine and Cogentin, stating she did not need them.

I was not psychotic, and according to Compendium of Pharmaceuticals and Specialties, precautions for the antipsychotic drug Loxapine, and the side effects and contraindications of that drug, I was indeed within my rights to refuse to take it as I believe that I should never have been prescribed it for the symptoms that I was experiencing.

[1]Loxapine is contraindicated with seizure disorders, as it may lower seizure threshold. It is responsible for CNS side effects such as sedation and drowsiness. Blocking alpha-adrenergic receptors can cause orthostatic hypotension and reflex tachycardia.

These were the symptoms that I was experiencing when I was admitted to the hospital, while I was taking Epival.

Patient granted weekend pass unaccompanied. Patient remains grandiose. Left on weekend pass with medications and instructions.

Patient informed to return to unit January 18, 1997 at 2100. Patient appeared receptive to same.

Sunday, January 18 2100

The weekend went well,
When I returned to St. Clare's Hospital, I ran into my cousin's wife by the main entrance. She informed me that my Aunt Mary was hospitalized to have a knee replacement in the morning. I went up to her room to visit her.

Psychiatric Nurse's Notes
January 18, 1998

Returned from pass on time, looked bright and in good spirits. Said pass had gone well. Spent time in lounge with other patients. Routine observation.

Monday January 19, 1998.

I got up early and went down to the operating room to see Aunt Mary before she went to have her surgery.

I spoke to the anesthetist, whom I knew and inquired about my aunt and asked him what type of anesthetic he would be giving her. He informed me it would be an epidural, as she was ninety years old. The nurses there, whom I knew very well, told me I could come back and stay with Aunt Mary as soon as she came out of the operating room as most likely she would be fully awake, but she would have to stay there for at least an hour.

I was not allowed to go back to the recovery room and stay with Aunt Mary as I had been invited to do, and could not understand why. This made me upset, as I knew Aunt Mary would be expecting me to be there, and I had no way to explain to her why I was not. I was being badgered again.

I was informed that I was not being discharged, after my case was discussed at rounds. I could not understand that either.

I tried to contact Dr. Frecker, and I was unsuccessful.

One of the nurses became very nasty towards me and was lecturing and scolding me until I thought I would lose control, but finally, after listening to her as long as I possibly could, I had no other choice but to ask her nicely to "just shut up" twice before she actually did and then I walked away from her, which I really should have done earlier.

After that bullying, I checked myself out of hospital again before I would get into the same situation as I did at the Waterford Hospital.

---

[1] Compendium of Pharmaceuticals and Specialties, CPS 2008, Loxapine precautions pg. 1276.

After I came home, I wrote this nurse a letter and apologized to her for saying "shut up," but I had to try to make her realize that if she treated patients the way she had treated me, she was most likely making matters worse for them.

Psychiatric Nursing Notes
January 19, 1998

0800-1215

Up and dressed, awaiting a visit from Dr. Frecker as she has four pages of questions for him. Patient encouraged to go over these questions with resident.

I wrote down questions on a notepad that I wanted to discuss with Dr. Frecker, but I did not get the chance to do that as I discharged myself for fear of getting certified again and being forced to take more of these psychotropic drugs that I felt I did not need. By now I really believed they were making me worse and could possibly kill me.

Below are the 'four pages' of questions that were referred in the nurses' notes to which I had received no answers.

What is causing the numbness, tingling, paralysis upon awakening, which causes palpitations? Are these palpitations serious? Is the Epival causing this?

The treatment that works best for the pain and stiffness is taking a hot bath as I cannot tolerate pain medication and besides they don't help the pain and stiffness anyway. Medication makes me sicker. why?

What is causing my bruising to return? What can I do to stop that? Would it be the Epival causing the bruising?

[1]Later when I was studying these drugs I learned that According to CPS, bruising is a known side effect of Epival, as the drug alters the normal clotting mechanism. This serious side effect of bruising was overlooked by the physicians.

What about tea as a diuretic? I find the Diazide is too drastic and I get headaches. I have to restrict Diazide to just half a tablet once or twice a week.

Is this post traumatic stress from my past treatment especially at the Waterford Hospital?

Will my memory ever get any better?

How can I contact a support group? I would like to get involved in research and treatment regarding fibromyalgia, so I can react positively to this misunderstood chronic illness.

Since I have been diagnosed with fibromyalgia, I would like to cooperate with the physicians now so I can benefit to the fullest regarding my treatment towards a better prognosis. The past is buried. Let us forgive ourselves and each other.

I was really trying very hard to obtain some understanding regarding my medical condition and the symptoms that had led me to seek medical help for the third time this month, but it seemed I was going deeper into the psychiatric trap. Next to the Waterford Hospital stay, It was the most stressful time of my life.

My mother wrote a poem on Forgiveness that I want to add here.

Forgiveness

I've found a little remedy
To ease the life I live
And make each day a happier one
It is the word forgive

---

[1] Compendium of Pharmaceuticals and Specialties, 2008 Epival, warnings pg. 818

Sometimes some little things come up
That give a pain or sting,
But if covered up at once
They won't amount to anything.

But if I keep them in full view
And pout and moan and fret,
They grow much bigger every day
Just forgive and then forget.

Nurse's Progress Notes
January 19, 1998
1050

Patient was discussed in rounds but not discharged as she wished. Informed that she could see Dr. Kennedy if she insisted on signing out against medical advice, but it would be around one hour.

Patient proceeded to page Dr. Frecker overhead, telling the operator she was a nurse. Dr. Frecker did not answer page. (This was definitely written out of context, the operator recognized my voice. (I was not out of order and I behaved in a professional manner as I always had.)

She was asking to go into the Operating Room to visit her Aunt as she was "invited" due to her position (Nurse at Health Science Centre). Operating Room confirmed this was not true. (This was indeed true.)

Patient very irritated telling writer to "shut up" times two.

Patient left unit. Attempting to see Dr. Ladha. Again advised this is not appropriate (by Dr. Ladha's secretary)

Patient then explained she had an appointment with Dr. Doucet tomorrow.

She was seen by resident and discharged against medical advice.

This writer bullied and badgered me. Then she wrote the above progress note to demonstrate that I acted inappropriately, when indeed I had not. I was mentally stable and behaving appropriately, but she was controlling me and had tormented me unnecessarily. After being diagnosed with fm/cfs/mcs, I believe that I should not have been admitted to the psychiatric floor during my last presentation to emergency at all.

Resident Note
January 19, 1998

Judi seen today 1155-1210 requesting discharge immediately.
Seeing Dr. Frecker this morning patient requested discharge.
She became increasingly upset with nursing staff.
Now requesting discharge against medical advice.
She states she is feeling well, no thoughts of self harm or harm to others. She had good weekend pass without problems.
She has difficulty with hospitalization especially rules regarding bath times. Compliant with Epival. (This was inaccurate. I had refused Epival since I had been admitted as it had caused me enough problems.)
Insight remains poor and denies bipolar diagnosis, claiming problems are secondary to fibromyalgia.

Cooperative, good eye contact. Speech spontaneous. Mood "fine but irritated" affect congruent with mood.
Thoughts –no perceptual/ form disorders.
Attention/concentration –slightly decreased.
Orientated times three.

<u>Impression:</u> **Not certifiable**

<u>Plan:</u> **Discharge against medical advice-Dr. Frecker aware**
   Appointment with Dr. Doucet tomorrow.

My memory and concentration were severely decreased and if not for my notes that I kept writing to jog my memory, many of these incidents would never have entered into my memory bank, even though I remember distinctly this aggressive nurse bullying me and the argument I had with her as she was far from being professional in her dealings with me. Her attitude towards me and the fear of certification initiated my decision to discharge myself against medical advice as I thought I would end up back into the Waterford Hospital again and that would be the end of my life.

This is a summary of my symptoms for the past year.

February 18, 1997

- Rash on my neck to my breastbone.
- Anxiety and panic attacks for the first time in my life.
- Started having difficulty voiding.
- Indigestion problems.
- Feet sweating and burning, calming myself in the bathtub.

February 28, 1997

- Itching all over and burning skin.
- Reddened areas on legs and arms which were hot to touch.
- Shivering and burning at the same time, after taking Ativan

September 24, 1997

- Forced to go to hospital by ambulance, while police attended.
- Emotionally tortured for six hours in an emergency room.

January 9, 1998

- Increasing muscle stiffness and weakness
- Extremely exhausted and physically weak
- Palpitations,
- Chest pains
- Shortness of breath and anxiety after taking Haldol with Epival.

January 12, 1998

- Numbness intermittently and tingling sensations
- Paralysis at times upon awakening,
- Pains in chest, between shoulder blades
- Shortness of breath and palpitations.
- Painful hips. Could not lie on my sides.
- Heightened sensitivity to everything.
- Difficulty in getting from a lying position to sitting or standing.
- Loss of balance, and blood pressure dropping while standing.
- Slow moving, dragging left leg and shuffling instead of walking.
- Difficulty climbing stairs.
- Difficulty swallowing and digesting food.
- Bruising
- Memory impairment
- Itching all over

These symptoms had progressed over the past two years, but since I have been taking Epival for three months, I was much worse and felt now that I was slowly dying.

January 14, 1998.

I was brought by ambulance to emergency because of waking up with palpitations, shortness of breath and pain between my shoulders and not enough muscle strength to get to a phone to call for help.

By this time I had been reduced further to a 'babbling' idiot and these drug induced, physical symptoms were overlooked, and the diagnosis of bipolar disorder was not revisited.

January 19, 1998

I discharged myself against medical advice for fear of recertification and being forced to take drugs that were making me worse.

The final discharge diagnosis was written with a question mark in front of it. (? bipolar disorder)

In a sick room, ten cents worth of human understanding equals ten dollars worth of medical science. / Martin H. Fischer

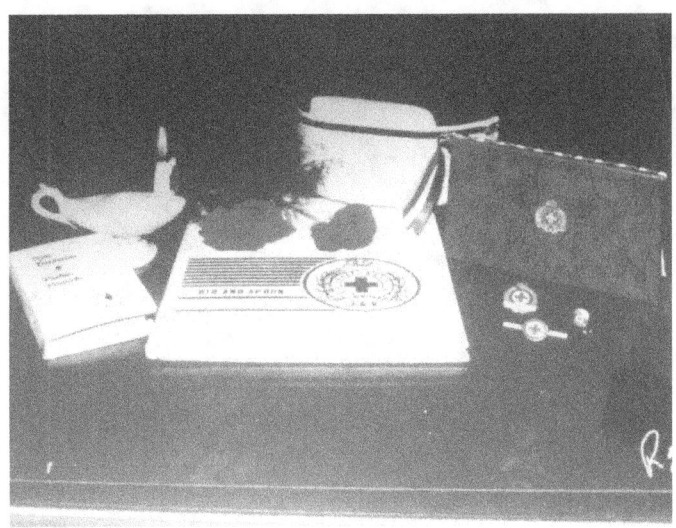

My graduation memorabilia from Grace Hospital School of Nursing where I was taught the Art, Science and Spirit of Nursing

Chapter 21

Finally a Diagnosis of CFS/FM/MCS

My daughter Karen phoned me after I arrived home from the hospital as apparently my brother Jim had contacted her while she was in Florida visiting her brother Brian, who had moved there with Sabrina. He was working there, managing a restaurant.

Jim said to Karen, "Your mother is going to end up back in the Waterford Hospital if she does not take her medication." Karen informed me that Brian became very upset because of the experience he had the last time, when nobody would listen to him when he was telling the doctors that I was fine and to let me go home.

Karen told me Brian was crying out of frustration. I reassured Karen I was fine, but I do not know how to explain to the doctors that I am not bipolar. I have fibromyalgia and chronic fatigue syndrome. All the stress that I am being inflicted with, due to having no control over my treatments, is making me worse. I don't know how I will get out of this psychiatric trap, but I will keep trying.

I slept well the night I came home from the hospital and kept my appointment with Dr. Doucet the next day. By then, I had not taken any Epival for four days and I decided that I was not taking any more as I believed that the Epival was causing me to have more pain, stiffness and weakness and I felt it may even be causing me to have the chest pain and palpitations that were waking me up. I had decided not to tell anybody that I had stopped taking Epival.

Dr. Doucet's Consult

January 20, 1998 1330-1350
Bipolar Disorder
    Discharged yesterday after one week.
    Was on Epival 250 mg BID (twice a day) for 4 days. Volporic Acid level January 19 4/4
    CBC January 19th Platelets 259.
    Sleep--- good
    Energy--- good
    Pains and stiffness- paraesthenia (muscle weakness trunk and limbs.
    Has noticed some bruising in her legs.
    Memory poor. Distractibile.
    She was a little over anxious and pressured but ability to stay on track fairly well.
    Plan: continue Epival 250 mg Twice a day (illegible) blood work.

Even though I had not taken Epival for four days, my blood still showed that the level was 4/4.

For the next month I had no more palpitations upon awakening and was coping better with my health, but pushing myself to try to regain my strength and energy.

With the help of Margaret whom I had hired for a few weeks, we straightened out my closets and my home to make it as relaxing and therapeutic as it could possibly be. We set up a gym in one of the bedrooms and I bought mirrors to place on the walls to make the room look bigger and brighter. I had a stair climber, exercise bike, rowing machine and other smaller pieces of exercise equipment that I had acquired over the years.

I was exercising as much as I possibly could in my gym at home, but would become exhausted after being on my exercise bike, stair climber, rowing machine, or abdominal machine for less than a minute. I kept trying to increase those times as much as I possibly could, but my legs and arms would become so weakened then they would pain.

When I could not stand the pain any longer, I would have to stop. Then I would get on a different machine and kept up the exercises, until I had gone on all the equipment that I had in my gym. I did this routine at least twice a day.

Dr. Doucet's Consult
February 13, 1998 1410 -1500
Bipolar Disorder

Very stressed coping with legal issues of her past medical treatment. She feels she was misdiagnosed and mistreated from a medical point of view.

Discussed how this may have affected her but will review only issues pertaining to her health, current symptoms and diagnosis.

Felt traumatized by past certification and hospitalization at Waterford and also feels it has led to a mistrust of physicians.

She is pressured in speech and agitated by discussing above issues. Mood tense and irritable
Encouraged to increase Epival to 250 mg am and 500 mgms at bedtime.

After another week without medication, I visited Dr. Doucet again, but did not tell him that I had been off all medication for over a month. I was now digesting my food better to the point that I was having abdominal cramping.

Dr. Doucet's Consult
Feb 27, 1998 1300 - 1400
Bipolar Disorder

Had increased Epival to 250mg and 500mgs at bedtime, but only briefly. She notes her energy then was very poor, generalized pain and stiffness. Very poor exercise tolerance. Energy easily drained with any exertion.

Also complaining of epigastric and lower abdominal pain and cramping.

Sleep is increased to twelve hours a day.

Periodic problems with concentration.

Discussed again diagnosis and appropriate treatment. She does not agree with the diagnosis and fears a diagnosis of another illness may be missed because of our (psychiatry) bias in favor of bipolar disorder.

Discussed above issues at length.

She is a little pressured in speech, stays well on track in discussion, mood good, energy poor.

She was advised again to continue the Epival at 250 mg and 500 mg at bedtime.

After two months of not taking **any** Epival or any other drug, I finally told Dr. Doucet that I had stopped taking them. I have kept the last prescription that he had given me as proof that I did not get that prescription filled.

I also want to bring you to the attention that Dr. Doucet discontinued Epival when I was admitted to hospital, January 9, 1998 and the doctors were trying to get me to take Loxapine and I had refused to take it. Then I was prescribed Epival again, but did not consume it.

After I came home from St. Clare's Hospital and studied Epival and its adverse effects, I definitely believed it was the Epival that caused me to wake up with palpitations and shortness of breath and a feeling I was dying as I had become so weak.

These symptoms had prompted me to go to the Emergency Department, but I was just a psychiatric patient and again my physical complaints were dismissed as psychosomatic as I was diagnosed as bipolar-manic.

Since I have been off Epival, I have had no more signs of waking up with palpitations, shortness of breath or chest pain and the bruising had subsided again. I realized then that I had experienced major side effects of that drug and if I had kept taking it, I believe I would have had a cardiac arrest in my sleep as my muscles were so weakened by taking this drug that it had affected my heart.

I was blessed that I do not have a family history of heart disease or perhaps I would have died the same way my mother did with a drug-induced cardiac arrest.

Sometime in March, 1998, during one of our phone conversations, Karen and I agreed that I would get a consult with a lawyer and go through the courts via the Mental Health Act and Law. She offered to help me financially as we knew the process was going to be expensive.

I felt I was compelled to go this route as the illegal detention and misdiagnosis at the Waterford Hospital in 1995 will come back to haunt me the rest of my life. The lawyer's advice to go through the Newfoundland Medical Association had not accomplished anything. As far as the medical association was concerned I had been treated according to their medical standards and the doctors had done no wrong.

I wanted to be able to feel that I could go to an emergency department with physical complaints and be treated objectively like any other normal human being. I had now become scared of seeking treatment because of the threats of being admitted to the Waterford Hospital and certified under the Mental Health Act. Who will stop them from doing this now if I do not?

By now both my hips were very painful and I could not sleep on my sides, and I had a heightened sensitivity to everything. The muscle stiffness and pain had progressed to be all over, especially upon awakening to the point it was difficult for me to get out of bed in the mornings and it was very painful when I put my feet to the floor.

I felt I had to keep moving, keep going beyond the pain to do anything or move anywhere. Then as soon as I would sit down again, I had to experience all this pain again. I conditioned myself just to keep going beyond the pain, as by this time, I did not want to take any more medications, not even an aspirin.

I waited eight months to get a consult with a rheumatologist. In June 1998, the diagnosis of chronic fatigue syndrome/ fibromyalgia was confirmed by Dr. John Martin. This diagnosis confirmed what Dr. Rice and Dr. Button had determined and I believed that was my correct diagnosis. All I wanted was for the diagnosis of bipolar disorder to be erased from my records and the mistake acknowledged by the medical community so I could have my credibility restored, my life back, and be gainfully employed without having to take psychotropic medication that I just could not tolerate due to my chemical (drug) sensitivities.

I had learned by then that many people, before they are diagnosed with cfs/fm/mcs have been misdiagnosed with a psychiatric illness, mostly depression. They go on for years as I did, without receiving the proper diagnosis and treatment.

Many other people get mistakenly diagnosed with paranoid schizophrenia, and others like me, subsequently end up diagnosed with bipolar disorder. The treatment for bipolar disorder is very different than the treatment for cfs/fm/mcs and it makes the symptoms of fm/cfs/mcs worse.

According to Dr. Eleanor Stein These are the main differences between chronic fatigue syndrome and depression.

Chronic Fatigue Syndrome	**Depression**
1. Fatigue is the primary symptom	1. Mood change is primary symptom
2. Both physical and mental fatigue	2. Physical fatigue rare
3. Fatigue both physical and mental worsened by physical or mental exertion (Blackwood et al, 1998)	3. Fatigue and mood improve with exercise
4. Decreased positive affect (energy, enthusiasm, happiness)	4. Increased negative affect (apathy, hopelessness, suicidal ideation, self reproach
5. Attributional bias only for somatic complaints	5. Generalized negative attributional bias
6. Externalizing attributional style (Powel et al, 1990b)	6. Internalizing attributional style (Powel et al, 1990a)
7. Personality disorder no more common than control samples (Pepper et al, 1993)	7. Increased prevalence of personality disorder than controls (Pepper et al 1993)
8. Infectious onset in ^80% of cases	8. Rarely follows infectious illness
9. Physical (somatic) symptoms; requires at least 4. eg. swollen lymph nodes, sore throat, muscle and or joint pain and headache.	9. Not usually associated with physical symptoms.
10. Cognitive dysfunction most problematic symptom present in absence of depression (slow reaction time and slow effortful processing. (Crowe & Casey, 1999) (Michiels & Cluydts, 2001)	10 Similar to CFS but able to be differentiated on neuropsychological testing (Daly et al, 2001)
11. Orthostatic intolerance/autonomic dysfunction (Rowe & Calkins 1998)	11. No association with autonomic symptoms
12. Sleep disorder Common	12. Sleep disorder common
13. Diurnal (daytime) variation with pm the worst time of day	13 Diurnal (daytime) variation with am the worst time of day
14. Variability of severity and nature of symptoms	14. Symptom variability not marked.

The sine qua non of chronic fatigue syndrome is persistent fatigue plus at least four other physical symptoms.

The sine qua non of clinical depression is a persistent low or irritable mood and anhedonia (absence of pleasure), guilt or self blame.

In order to be diagnosed with CFS, a patient must have met the following criteria according to the Clinical working Case Definition, diagnostic, and Treatment Protocols

- **Fatigue:** new onset, unexplained recurrent physical and mental fatigue that significantly alters activity level.

- **Post –External Fatigue:** after physical activity there is an increase in symptoms and/or an extended recovery period usually lasting a day or more.

- **Sleep dysfunction:** There is unrefreshed sleep and/or difficulty getting to or maintaining sleep.

- **There is a significant degree of muscle pain:** Pain can also be experienced in the joints and is often wide spread and changes location. Often there is a new onset of headaches post-illness. Headaches may be a different quality and in a different location than in the past.

- **Two or more neurological/Cognitive Manifestations:** confusion; impairment of concentration and short term memory; disorientation; difficulty with information processing; categorizing, and word retrieval; and perceptual and sensory disturbances. The Expert Panel describes overload phenomena cognitive and sensory- i.e., heightened sensitivity to lights and noise- and/or emotional overload, which may lead to "crash" periods and/or emotional symptoms.

- **At least one symptom from two of the following categories:**

    Autonomic Manifestations: blood pressure abnormalities, particularly when rising from lying or seated position, often called delayed postural hypotension; light headedness; nausea and irritable bowel syndrome; urinary frequency and bladder dysfunction; heart palpitations; shortness of breath with physical activity.

    Neuroendocrine Manifestations: "Thermostat" regulation is lost, presenting as lowered body temperature with significant daily fluctuation, sweating episodes, recurrent feelings of feverishness and cold extremities; intolerance of extreme heat or cold; significant weight change- lack of abnormal appetite; worsening of symptoms with stress.

    o **Immune Manifestations:** Tender lymph nodes; recurrent sore throat; recurrent flulike symptoms; general malaise; new sensitivities to food, medication, and/or chemicals.

- **The illness is chronic and lasts at least six months in adults, three months in children. It usually has a distinct onset, although it may be gradual.**

(1) Carruthers et al., Journal of Chronic Fatigue Syndrome 11, no.1 (2003)

This is an article I had written on fibromyalgia in the late nineties and have revised it occasionally to add new information as it unfolds. Over a period of five years, 1998-2003, I gave a hand-out to hundreds of patients when I was leader of the Self Help Group of Newfoundland and Labrador.

During that time, Dr. Button had told me he felt that I was the most knowledgeable person in Newfoundland and Labrador on this illness as he was aware that I had spend countless hours of my time researching and educating others by spreading the information as it became available by mail, newspapers and radio throughout Newfoundland and Labrador.

FIBROMYALGIA IS REAL

Fibromyalgia, described as a chronic invisible illness, is an under diagnosed neuro-endocrine, central nervous system disorder of unknown origin affecting 2-4% of the general population, mainly women. It can affect people of any age, wounding them, but leaving no physical scars. It is called an invisible illness as people suffering with much pain, stiffness and fatigue appear healthy, and have been described as "Heart Sink Patients".

Presently fibromyalgia, chronic fatigue syndrome, and multiple chemical sensitivities, which usually go together are being taken more seriously, but funding for research is still lacking, therefore these illnesses are still taking a toll on people's quality of living.

Most recent research supports that there is a brain defect affecting the hypothalamic-pituitary-adrenal, (HPA) axis that leaves a person unusually challenged because of a hyperactive pituitary release and a relative hypoactive adrenal response, resulting in a disturbance in the reactivity of the sympathetic nervous system.

A brain injured by trauma, viruses, toxins, or severe emotional and physical stress seems to trigger the symptoms of Fibromyalgia. This abnormal physiology can be better understood now due to brain mapping, scanning and electroencephalograms, which can be done with new and improved technologies that are available to the medical and scientific community.

This defect of the brain and central nervous system plays havoc on all body systems, which causes malfunctioning in many organs and tissues. There are approximately twenty-five symptoms that sufferers could complain about, so people with fibromyalgia, before being diagnosed, have also been described as "physicians' nightmares".

Treating people suffering from fibromyalgia "piece-meal" is going to add more stress and insult to them, creating considerable physical, mental, spiritual, social and economic implications. I believe holistic therapies or treating people as whole beings will create more positive results.

Unfortunately many patients, approximately 65%, mostly women, get misdiagnosed as having a mental disorder, and told their symptoms are considered "all in their head," before being diagnosed properly.

Treating fibromyalgia with normal dosages of antidepressant, psychotropic, neuroleptic and other neuro-toxic medications will exacerbate the symptoms of fibromyalgia as studies reveal that patients with fibromyalgia are highly sensitive to these medications. Some can only tolerate one tenth of a normal dosage, without experiencing serious side effects that will mimic mental illness.

Fibromyalgia affects four women to every man, so many women have suffered both, mentally and physically, needlessly for many years before being diagnosed properly and treated.

After a diagnosis of fibromyalgia is established, patients must take charge and arm themselves with knowledge regarding reduction of physical as well as emotional stress, whether it be positive or negative, which can cause flare-ups. They must learn healthy lifestyles, balancing their activities, pacing themselves, self- care and therapies so they can manage their own condition in order to live a better quality of life. At present there is no cure, only treatments, mainly holistic to reduce symptoms. Above all they must remain positive,

Research is ongoing to find better ways to diagnose this condition. Different antibodies have been detected in the blood of many people with fm/cfs, one being APA, anti-polymer, (toxins) and other viral antibodies. Many post polio survivors are now experiencing fm/cfs symptoms, as it seems that the polio virus is rearing its ugly head again in their bodies and negatively affecting their damaged and repaired motor neurons some forty years after the onset of polio.

This past few years, many physicians and scientific researchers are taking a closer look at this debilitating disorder of FM/CFS. Now physicians who would dare say that it is not a real physiological disorder are not educating themselves with the latest research on these illnesses.

Hopefully, as research is done and progress is made, there will be better treatments or a cure found. There are new drugs Pregabalin, (Lyrica) and (Cymbalta) being prescribed now that help with the pain, but the side effects of those drugs are very disabling for most people. These drugs reduce the release of or increase specific brain chemicals that may cause more fatigue, side effects and negative symptoms than the condition itself. After my experience, I do not believe that medication is the answer for me. I work on diet, exercise, rest and relaxation techniques.

It is important that people with those conditions receive support, understanding and belief from all health care workers, insurance companies, family and friends in order for them to maintain the courage to cope with such a demoralizing and debilitating condition and still lead a quality productive life, which I believe is possible.

Written and revised periodically by Judith Day

Living with Fibromyalgia

To make it easier to understand, I personally describe fibromyalgia as a cousin to multiple sclerosis (MS). MS can be described as the electrical wiring system of the body being damaged, which destroy the insulation of the networks that causes breaks in the neuro-transmitter system,

With fibromyalgia, the whole transformer cuts out and in. This malfunction causes most of the same symptoms of MS, but more inconsistently. What's the good of a wiring system if the energy is not getting through?

This abnormality plays havoc on all our systems, causing malfunctioning in many organs and tissues affecting every system in our body. There are approximately twenty-five symptoms that sufferers could complain about, so people suffering from fibromyalgia, before being diagnosed, have been described as physicians' nightmares.

Dr. Devin Starlanyl, who has the condition, described fibromyalgia as a chronic invisible illness that inflicts terrible wounds on you, but they are wounds that leave no scars. I beg to differ, maybe no visible scars, but emotional ones,

Living with Fibromyalgia as a child

I will try to tell you what it was like living with fibromyalgia for most of my life and not knowing what was wrong.

At the age of two years I was tipped over in a wooden highchair and landed on the back of my head. This may have been the beginning of fibromyalgia for me. According to the MRI that was done in 1997, I was told there is a scar on my cerebellum.

Mother told me laughingly that I was the only child she knew who asked to go to bed as I would get so tired. I had bouts of losing my strength. Once the doctor thought I had polio, but I recovered without any deformity or paralysis. Many times I would find a spot, inside or outside the house and fall asleep during the day.

I was a very sensitive child and remember being very hurt when I was laughed at because I could be so sick and tired one day and by the afternoon of the next day fully recovered. I remember my mother saying "I was so worried about Judi when I left for St. John's. By the time I got home the next day, she was fine. I believe the new coat I bought her made her better." Very early in my life, I learned not to complain about the way I felt, because I was not taken seriously. I suffered in silence.

Despite my bouts of tiredness, and "growing pains" as it was called, I was considered a very pleasant, lovable and happy child who received from, and gave a great deal of affection to everyone.

I suffered terribly from nasal stuffiness, when I would lie down, especially during the summer. I constantly carried a Vick's inhaler. I remember the doctor telling my mother I had the sniffles. My nose was always runny.

I also suffered from severe headaches, over my right eye and through the right side of my head, especially on hot sunny days. I can cry now thinking about how much I suffered with these headaches. I would lie in a dark room for hours, become nauseated and vomit. Then I would feel better. When I could eat again I would ask for something with mustard on it.

I was motion sick whenever I travelled. It helped if I looked directly in one spot or lie down with my eyes closed. I thought and I would say that my eyes made me sick. It wasn't the motion, it was my eyes, as watching a movie or television would make me sick also. Sometimes looking at bold prints or plaids, or checks would make me nauseated too

I remember having blurred vision and abnormal blinking. I was taken to St. John's to see Dr. Lynch. By the time I was examined, I could see properly and I did not need glasses. He told my mother that I had some sort of inflammation behind my eyes, which he treated with drops. There are many times that images appeared brighter than other times. It is almost like when you turn on an extra light. I had a lot of sties on my eyelids. As soon as one got better another one would start. They were very painful and made me tired.

Colds and flues were a big part of my young life. I also had a bout of pleurisy. I can still feel the pain when I breathed. I was in bed a long time getting over that. Every cut or scrape meant infection. I can still remember my mother cleansing my sores with boiling salt water.

I would get terrible cramps in my stomach and without much warning I would have diarrhea. When I would run and play, if I did not twist my ankles, I would get "stitches" in my sides and would have to stop because I would get completely out of breath. I remember while walking through a Park one day, there was a loud noise behind me. I completely lost control of my legs and fell to the ground. My legs felt like rubber.

When I swam I would get leg cramps that would come on suddenly especially if I extended my leg too far or pointed my toes. I remember once I was far out in the pond. Thank God there was a shallow reef under me or I would have drowned, even though I was a good swimmer.

During the winters, I would get cold hands and feet easily. Once I had to go to the doctor because my toes were swollen and red. He told my mother that I had chill blains. My hands and feet would get cold, go numb, and then they would feel like they were burning.

The headaches started to be less severe, but with menarche came fainting spells, rashes on my armpits, no acne though, but mouth ulcers and cold sores galore. My hair became very oily. I had to wash it daily.

As a teenager, I still required a lot of rest and I went to bed very early at night. I could not stay up late and function the next day. Rarely did I over-tire myself, because that meant headaches. I remember once after a high school dance, I was in bed all weekend because my legs were so weak and sore. When I participated in sports, I would have pains in my legs and could not understand why nobody else playing the sport felt like I did. The soreness would go away, if I got past the first few minutes of exercise. I learned to live with muscle pain as just being normal. Very early in my life I learned that pain could be "worked off or exercised away." I tried not to complain too much as my mother was not well herself and I felt I was adding too much to her worries.

She kept taking me to the doctor, but nothing seriously was wrong with me.

Looking back over my life, I believe that I would have had a much better quality of life if only I had known that something physically was wrong with me. I would become so frustrated many times because of how sick and tired I felt and not knowing why. I thank God that I did not lose my spirit or zest for life and despite Fibromyalgia I have become the person who I am today.

Living with Fibromyalgia as an Adult

When I entered nursing at age 18, I was diagnosed with walking pneumonia during my routine physical exam. I had neither temperature nor symptoms except fatigue and a headache, which was very common for me to have. I was placed on bed rest for a week with antibiotics. It took me about a month to get over that bout.

After that it was one attack of tonsillitis after another, mouth ulcers and cold sores. Sometimes I would stay off work because my lips were swollen so much, I looked deformed.

I started having mucous in my stools shortly after that and a barium enema show no evidence of anything. By now I had learned the term "hypochondria" I felt now that I am turning into a hypochondriac and I had to stop complaining about things because nothing is really wrong.

I kept being involved in sports and swam a lot. When I felt sick or tired I would stay off duty in bed, I felt guilty because my temperature was always "normal." My normal temperature is around 96. I used to think that people thought I was faking.

I managed to complete my nursing by mostly working in the operating room during my last year, where the environment was more controlled. I only had to look after one patient at a time so I did not have to worry about others who I was responsible for when they were out of my sight. If I had not found the operating room, I would have had to leave my nursing career because I could not physically keep up. It was affecting my health.

I remember being reprimanded by a supervisor because she caught me standing up, leaning over with an infant propped up in my arms, on a change table, because my back was paining so much from sitting on stools while feeding them. Again I suffered in silence.

After graduation from nursing school, my first job was working in the operating room at Labrador City. We had to work a lot of overtime. I became exhausted and not knowing why, I just told the supervisor one day that I can't work like this anymore and that I would have to quit. I went home and rested in bed for about three weeks, before I went back to work again. I would have to take naps, especially if I were up the night before on an emergency. I could not physically keep going.

My physical problems were attributed to the birth control pills I was taking, so I was advised to stop taking them. I had abnormally long and difficult deliveries for both my children.

My nursing career was mainly part time as I could not physically keep up with full time work. When I had to work full time, for financial reasons, I always had a live-in homemaker to help me until my children were grown big enough to help me.

Every holiday I went on was ruined because I would become sick and tired after the plane ride. I would rather lie on the beach or in bed than go on a tour. I would have to return home at least three days to rest before going back to work. I still had frequent bouts of infections. I had bronchitis five times one year. They would come on suddenly, overnight. The fatigue would last for weeks.

My social life has always been excellent and most of the time I feel very contented.

In 1976, I was involved in a seven car pileup going through a tunnel in Montreal. I cannot say my symptoms got any worse after that accident. They were the same. I was off work six weeks recovering from head and knee injuries.

Until 1995, I thought I had lived fairly normally with fibromyalgia and coped very well living with the symptoms. Of course my positive attitude helped. There is documentation on my medical records confirming I have had every symptom of fibromyalgia that has been researched and recorded. I can honestly say I knew nothing about fibromyalgia until 1997.

Drug sensitivities and adverse reactions are very common with fibromyalgia. Research shows that doses of one quarter to one tenth of what a normal patient can take are not uncommon. People with fibromyalgia should be started on the smallest amount possible and increased slowly.

I have had a hard struggle living with fibromyalgia. I am writing this book to help me deal with that struggle. It is one of my resources for healing my emotional scars and hopefully others can relate to what it is like being misunderstood, undiagnosed and then worst of all misdiagnosed with manic depression and treated harshly, unfairly and without respect for almost three years, until I managed to get out of the psychiatric trap and finally in 1997, get diagnosed properly as having CFS and fibromyalgia.

My mentor and hero, Dr. Devin Starlanyl, who also has fibromyalgia, writes "Even when you are tarred and feathered you should still try and fly."

Since I have been diagnosed and understand the illness, my quality of life has improved. I know now why I am like I am and listen to my body while I take good care of myself.

> I became stronger than I could ever imagine.
> I am more compassionate than ever before
> I now know who my true friends are.
> I have discovered my life's real purpose
> I know I have endless opportunities
> I have more balance in my life, and definitely have more patience.
> I am nourished by living simply, loving generously, caring deeply,
> speaking kindly, and leaving the rest up to God

Written by Judi Day, Fredericton, NB. 1998, revised, November 10, 2011

Making the proper diagnosis is vital: Statistics from studies 2002-2003 revealed that a substantial number of Canadians report symptoms of conditions that cannot be definitely identified through physical examination or medical testing.

These were known as "medically unexplained physical symptoms and they characterize conditions such as chronic fatigue syndrome, fibromyalgia and/ or multiple chemical sensitivities. **Health Reports, vol. 18, no 1**

The "Medically Unexplained Physical Symptoms" study reveals that in 2003, 5% of Canadians aged 12 or older, an estimated 1.2 million people, had at least one of the conditions listed below."

1.3% reported Chronic Fatigue syndrome. (They experience extreme tiredness).

1.5% reported Fibromyalgia. (They have pain lasting three or four months or more, in at least 11 to 18 specific areas).

2.4% reported multiple chemical Sensitivities. (They have a variety of symptoms, when they are exposed to synthetic chemicals in doses that usually have no noticeable effect).

14% of these people had more than one of these conditions; some have all three, as unfortunately I do.

According to the Canadian Community Health Survey, the percentage of women with each of those conditions was about double that of men, and the prevalence of mental disorders were particularly common among people reporting chronic fatigue syndrome.

These chronically ill people also sought assistance from both conventional and alternative health care providers and are three times more likely to have frequent consultations with their physicians more than ten times a year. One quarter of these people are likely to need home care and other assistance for daily living. That is four times greater than people who did not report those specific chronic illnesses.

The 2010 findings of the Community health Survey reported that the number of people reporting a diagnosis of cfs/fm and/or mcs increased by 25% jumping from 4.2% of the target population in 2005 to 4.9% in 2010. This increase of incidence of these illnesses has to be taken seriously by governments and the medical community as what really is going wrong with our people? Why are post SAR's patients coming down with the same symptoms? Should we be sounding bigger alarm bells?

If you would like to learn more about those illnesses in a straight forward easy manner, **Hope and help for Chronic Fatigue Syndrome and Fibromyalgia, Second edition by Alison c Bested, et al** is a wonderful book to add to your library.

After I studied my mother's medical charts from Old Perlican Hospital, I believe that my mother also had fm/cfs/mcs. I realized that she had the same symptoms that I have had throughout her life. I realize now that she, being the intelligent, strong and positive lady she was, learned how to deal with her invisible illness of chronic fatigue syndrome, multiple chemical sensitivities that exacerbated to the severe muscle pain in her body in 1983, after her bout with cancer, her bowel resection that resulted in infection and subsequently led to her having a colostomy.

In 1983, her large bowel came apart where it had been sewn together after the cancerous bowel was removed, spilling the entire bowel contents into her abdominal cavity and causing peritonitis and septicemia. (Infection throughout her entire body)

My mother became cold, pale and clammy to the point of being mottled. I did not think she would ever survive, but the body has a wonderful way of healing itself, when there is a will to live, and she always had that will, as I do.

A week after her bowel resection, she was taken back to the operating room where the surgeons opened her again and suctioned out liters of purulent fluid, rejoined her large bowel, and performed a temporary colostomy, which the surgeons placed high on her right side, probably thinking she would never live to care for it anyway, but she was a survivor and she lived.

Because she was a diabetic, the surgeons were reluctant to reverse that temporary colostomy, but it was such a nuisance for her to care for because it was situated just below her rib cage, making it very difficult for her to manage it. She kept asking me for advice whether she should take the chance of having it reversed or not. Knowing the risk factors and her chance of survival, I told her I could not advise her, as it had to be her decision alone, but I would support her, whatever that decision would be.

Finally, after a couple of years dealing with the colostomy, my mother could not cope with it any longer and said that she would rather die than have to deal with it, so again, she came to hospital for the surgery against all medical advice. But fortunately, she had private duty nursing care, the love and support of her family and friends, and the will to live, which made all the difference with her healing.

Throughout her lifetime, my mother like all human beings became anxious or sad when she had to face a crisis, loss or upset, but in her case, because of her metabolic /neurological disorders, it affected her physically with a rise in her blood sugars, a rise in her blood pressure, and a rise in her heartbeat. However she had learned how to control herself emotionally to prevent those physiological symptoms from controlling her and I have taught myself to do the same.

I observed my mother many times and have learned from her actions how to control myself during major upsets in my life. I noticed she would excuse herself and retire to another room if there were too much activity, excitement surrounding heated debates, for instance on politics, etc., even excitement and laughter would affect her. She rarely watched action packed movies or loud entertainment and kept herself calm and serene, as otherwise her blood pressure and blood sugars would rise and she would become physically ill and possibly have another stroke or heart attack.

It was only in severe circumstances, like when I was locked up in the Waterford Hospital or when she was humiliated by the incident in Sibley's cove, that it affected Mom's health to the point she was hospitalized with symptoms of congestive heart failure, (her EKG was normal) which were due to anxiety from those stressful events.

During my admission to the Waterford Hospital in 1995, this information was written on her medical record. "Worried about her daughter in the Waterford." Neither she nor I could have done anything about that situation.

I have learned so much from my mother regarding how to deal with this chronic illness that I keep writing about her, as I believe that she also lived with this disorder throughout her lifetime and managed it so well. Perhaps her own mother, my grandmother was also inflicted with fm/cfs/mcs, as she died from infection, six weeks after my mother was born.

My mother told me the stories about her mother and grandmother and wrote some of this information in her "Legacy."

When my maternal grandmother, Elizabeth Mary (Burden) Burry was six months pregnant, with my mother, her youngest son Otto, who was nine months old, died from whooping cough. After the death of her baby Otto, she seemed to have lost her spirit. When she gave birth to my mother, she weighed just four pounds. My Grandmother's sister Aunt Clara, (Gary and Susan Davis' grandmother) came to look after her and the family.

My Grandmother seemed to be slightly improving during the first month after my mother's birth, while my mother was steadily weakening. When my mother was less than one month old, she came down with pneumonia and was wrapped in linseed meal poultices for more than a week and finally started to recover.

At this time, my Grandmother was stressed worrying about her infant daughter and perhaps thinking she was going to die like Otto, had a setback with infection and milk fever, which took her back to her bed. Dr. Mac Donald from Salvage, Dr. Fraser from St. John's and Dr. Mahoney from Grand falls were called, but they could not save her and she died on September 3, 1911, leaving my mother just five weeks old, her brothers, Max and James, seven and eight years old and her husband Captain Baxter Burry.

This history repeated itself. When my mother was eight months pregnant, my sister Patsy, who was her first baby died suddenly when she was not quite two years old. My mother told me many times about that terrible time in her life, and how strong she had to be to protect and deliver her new baby, while struggling with the grief of the loss of her first born.

She told me that in her family history, both her mother and grandmother had died from child birth in their mid thirties. She was determined that she would survive, which she did, having two more children, my brother and me within the next two and a half years.

When my mother was pregnant with me she developed toxemia of pregnancy, gained a considerable amount of weight and Doctor Coyle placed her on bed rest for the last three months of her pregnancy. With two children who needed care and attention, which she was incapable of giving in her condition, she chose to send the baby Jim to my Aunt Mary in Glovertown. My two year old sister Betty was also taken care of outside the home.

My mother was intelligent enough to know that the demand on her of caring for these two children in her fragile physical condition was impossible. She had no other choice, if she wanted to save her unborn child, me, and perhaps herself. Maybe that was why we became so close, she had protected and nurtured me for three months even before I was born.

Jim stayed in Glovertown with Aunt Mary and Uncle Max for another three months until my mother recovered from my birth and she was well enough to take care of all three of us.

Dr. Coyle warned her not to have any more children, which was a hard thing to do in the forties and fifties, without the birth control methods of today. Therefore, five years after I was born and at the age of thirty nine and forty my two younger brothers were born eleven months apart. My mother told me years later, due to medical reasons, she had a tubal ligation, perhaps the first one performed in Newfoundland by Dr. Allan Wilkinson.

He probably saved her life as she could not physically keep up to the demands of caring for two babies, and being so fertile, having the possibility of becoming pregnant again.

Fortunately, my Aunt Belle and Uncle Charles, who had no children, decided to help mom take care of the two baby boys. Ford, the oldest baby was Aunt Belle's favorite and he spent a lot of time with them. Aunt Bell wanted to adopt him, but Mom could not bear to do that, so she made it quite clear that both the boys had to be treated alike and there had to be no favoritism with gift giving or anything else as they were so close together in age. Mom always ensured that everything had to be identical for them.

When the boys were old enough to go to school, my mother wanted to return to her teaching profession. She started back teaching school in New Melbourne, taking the two boys with her and taught them until they went to the Edwin J. Pratt Central High School, where Betty, Jim and I had already finished and gone away in pursuit of our careers.

My mother struggled with chronic fatigue and I still remember that. Betty, Jim and I helped her as much as we could with the daily chores. Each night when Mom and Dad came home from school, (Dad drove the school bus), we took turns having dinner cooked and afterwards, while mom rested in bed for awhile, we did the dishes and cleaned up before watching the evening news together.

Mom would then tutor our cousin Don and complete her home/school work for the next day. I learned from my mother that a teacher's job does not end when you leave the classroom at the end of the day, but it is one of the most rewarding careers that one can chose in life.

Mom hired cleaning ladies to do the heavy house cleaning, but otherwise Betty, Jim and I did the remainder of what had to be done on a daily basis. Dad was quite busy as well, with my Brother Jim's help, keeping the bus and other equipment in good repair and doing the other manual chores that needed to be done. My family was also involved in the fishery.

My mother was not physically able to keep going, especially with the fishery. She justified her fatigued existence by contributing to the family income by teaching, thus could afford to pay other people to do the physical work for her.

Perhaps people who knew no better considered my mother lazy, slow or a malingerer, but my mother was neither. I believe she was battling undiagnosed chronic fatigue syndrome, but despite coping with this illness, she led an interesting, contented and productive life as I have learned from her to do as well.

I can honestly say Mom and Dad did the very best they possibly could to provide a good and stable home for us. We thrived in their supportive environment and we learned how to become independent, motivated and industrious adults. A healthy and stable family life is something I have always cherished and felt it was a very important aspect of one's life.

I had a stable family life while I was growing up thanks to my parents, my extended families and friends. We had our ups and downs and arguments like most families, but I have to say, I had always felt secure and loved until the beginning of my marriage breakdown in 1973. I believe a happy family life and good family relationships are the essence of being healthy.

I, like my mother and grandmother also had difficulty with my pregnancies, and deliveries. During my second pregnancy, I suffered from nasal stuffiness for the entire nine months. By the time Brian was born, my body shut down and I was kept in hospital for ten days. Even Karen, while visiting me in hospital, who was just two years old at the time noticed how tired I was and said "you better go back to bed Mommy because you really tired." I did not suffer from post partum depression, just profound fatigue and a digestive system that was shut down to the point where I became impacted.

After I had Karen and Brian, I knew I could not physically go through another pregnancy and delivery, but had difficulty taking birth control pills, so I had three more pregnancies that ended in spontaneous abortions, all before I was thirty five years old.

I revised my mother's poem after I was diagnosed with cfs/fm/mcs by adding more verses.

Please keep me sweet as I grow old

Please keep me sweet as I grow old
Let my heart be forgiving
When unkind words are told.
Keep my mind full of good thoughts
So I can smile.
I want to forget things
That aren't worthwhile.

Invisible changes
Why can't people see?
What they've done to my body,
So ungracefully.
They've changed me forever.
I accept who I am.
But please understand me
I don't deserve blame.

Please keep me sweet as I grow old
Let my eyes keep on seeing,
Though the print be so bold.
Each day I grow older,
I don't want to complain.
I'll keep being positive
Though my limbs are in pain.

Please keep me sweet as I grow old
Put strength in this body

Make me straight when I fold
When this life is over,
And I'm ready to rest.
Thank you for helping,
Make my life the best.

Drugs are not always necessary. Belief in Recovery is / Norman Cousins

A full page article published by the Telegram on Fibromyalgia in 1999

Chapter 22

Taking my Life Back

I kept exercising in my gym at home until I felt I had enough stamina, and had progressed enough physically to go back to a real gym. When spring came, I joined the Aquarena gym and pool, and went there almost every day, beginning with water fitness exercises and stretching in the hot tub.

I wrote more letters of complaints to Dr. Young regarding the threatening treatment and the verbal abuse that I had received from Dr. Singh, and Dr. Ladha during the January admission to St. Clare's Hospital, as I felt that both their attitudes were uncaring and inhumane. Dr. Ladha, without giving me a chance to tell him what was wrong with me, had threatened to have me certified to the Waterford Hospital. I believed if that happened once before, it could possibly happen again.

I kept going to see Dr. Doucet, as it had been stipulated by Blue Cross that in order to continue with long term disability payments, that every two months, I would have to be followed by a psychiatrist, since the medical information on my medical files, which I tried to have corrected but was unsuccessful, does indicate an underlying psychiatric illness.

Consult Dr. Doucet
March 24, 1998 1705 -1758
Bipolar Disorder.

> **She has discontinued the Epival.**
> **She refuses to believe diagnosis.**
> **She believes all her symptoms are due to fibromyalgia or some other illness not yet diagnosed.**
> **She is complaining of fatigue and overwhelming weakness.**
> **At times she has episodic concentration and memory problems. Fluctuating muscular pain and weakness and off balance. She is staying active but limiting exercise to brief episodes. Resting well.**
> **She denies mood disturbances.**
> **She was advised to restart Epival or dangers of manic or hypomanic episodes.**
> **She also has abnormal EEG and should be on Epival.**
> **She has no insight into this and refused medication.**

I felt very worried and so alone as I had nobody who really understood what had happened to me, especially when I had shortness of breath, palpitations and then chest pain and just too weak to roll over in the bed, let alone get out of it.

For three months, I have had no more of these attacks since I stopped taking Epival. It was a relief and blessing to have the diagnosis of fibromyalgia and Chronic Fatigue syndrome confirmed by Dr. John Martin

I have been treating myself holistically since I discharged myself from hospital in January 1998 by:

- Taking vitamins, minerals and other supplements. I bought a juicer and make nutritious fruit and vegetable drinks with whey protein. I drink apple cider vinegar, green tea and eat other healthy foods that were recommended to detoxify myself and treat this condition. I keep myself well hydrated, which helps with the headaches.

- Using Tri-Salts. I took tri-salts for awhile, very sparingly, when I felt really ill, which helped me greatly and kept headaches away.

- Tri-salts are made by combining 3 parts sodium bicarbonate with 2 parts potassium bicarbonate and one part calcium carbonate, and made up in special pharmacies. I took only 1/8 teaspoon at a time and ensured I took no more than 1 teaspoonful in a week. It certainly helped me get rid of that flu like feeling and nausea that I had suffered for some time.

- Trying to avoid all stressful and tiring situations, whether good or bad until I became stronger. I have to balance my activities and rest periods and ensure I sleep well at night.

- Exercising as much as I am physically able to do each day by just going beyond the stiffness and pain, and trying to be as active both physically and mentally as I possibly can.

- Keeping myself well groomed in order to look the very best, and this helps me feel better. I believe I am feeling as well as I could expect to feel.

- Knowing and believing in myself that I would feel better and I do.

- Remaining positive, as I am confident that my suffering will not be in vain.

Three years to the month had passed since I was admitted and committed to the Waterford Hospital for five weeks with a diagnosis of bipolar, manic and discharged June 19, 1995. These years have been a long journey for me, but I have come out of it a much wiser person, with a better understanding of human nature and to what extent people would try to cover up for themselves and their colleagues, instead of admitting they could possibly be mistaken.

Now the ending appears sad, but this is not the end, this is the beginning of another journey of a new life for me, but first I had to work on regaining my strength and energy and try to feel better.

Somehow I believed that Dr. Doucet may finally be convinced that I was misdiagnosed, but that was not the case. During each visit, Dr. Doucet kept writing on my medical record the diagnosis "Bipolar Disorder" even though I have been taking no medication to treat that illness for over six months now and have not had any symptoms of depression, hypomania or mania and no more palpitations and pain between my shoulder blades.

Consult Dr. Doucet
June 25, 1998 1245 -1332
Bipolar Disorder

> **She has seen Dr. Martin, Rheumatologist.**
> **She was told she has diagnosis of fibromyalgia.**
> **Energy variation. Usually low tolerance, becoming fatigued easily drained after brief period.**
> **Attention, concentration and memory are poor. They may be good for brief period, about one hour at a time, but variable.**
> **She is trying to be physically active with some exercise on a daily basis.**
> **She is off Epival past several months. She feels more drained and tired on it.**
> **Acknowledged to her this is possible, but pointed out to her the dangers of manic symptoms without the mood stabilizer. She does not accept that she was ever manic or hypomanic.**
> **Plan. See on monthly basis.**

I had a few attacks of pain in my head and then one day during the summer, I woke up from a deep sleep with a very sharp piercing pain in the right side of my head, which lasted less than a minute. I had a male boarder living with me at that time and he encouraged me to go to the emergency department and get checked. I reluctantly went with him to the Health Science Centre after he reassured me that he would be my advocate.

I was seen by Dr. Patel, and for the first time in three years, I was treated like a human being. Dr. Patel, knowing me from working at the hospital, and the same physician who saw me when I went there the previous January, showed me his clip board, which was blank and said "see Judi it is a clean slate." I told him what had happened, he listened and examined me. Finally, I felt I was being treated like a credible patient.

I was given the proper neurological testing, and reassured that everything seemed to be fine and I was released from emergency very relieved that I was not having a stroke or aneurism. I was advised just to take it easy for a few days and come back if there were any more symptoms. I had no more severe episodes of pain, and in a few days went on with my activities as usual.

I had another scheduled appointment to see Dr. Doucet, which I kept.

Consult Dr. Doucet
August 17, 1998 1710-1758

She had few episodes of right frontal pain, which was present upon awaking. Made her feel weak, seen in emergency.

Frequent aching and pressure right frontal area and she is booked for repeat MRI in September.

Energy is chronically poor. She is exhausted easily by any physical or mental exertion. She is better able to "pace" herself.

She tries to follow a good lifestyle of diet and exercise and rest.

She feels generally weak most days and this is variable.

She continues off all medication

Cautioned about possibility of manic symptoms, but she does not accept that she ever had these symptoms.

She feels she was stressed and frightened and or physically ill.

Reviewed again for her the symptoms which were present and on which I make the diagnosis.

I acknowledge with her that she may have Fibromyalgia and also it is important to not miss any other ongoing illness.

She is today subjectively tired. She complains of problems of poor memory and fatigue. She looks well. She stops on track in conversation.

She is able to tolerate difference in opinion in diagnosis.

Plan see in 1//2 (6 months)

In the fall of 1998, I hired a lawyer to represent me in a medical malpractice lawsuit. Karen and I had decided in the spring that I had no other choice but to go the legal route with this medical negligence causing harm and nothing had changed since then. My lawyer filed a statement of Claim with the court November 23, 1998 and she sent interrogatories to the defendants and searched to obtain an expert medical psychiatric opinion outside the province.

We decided that I would try to get a medical expert from Fredericton to back up my claim, as I could come and visit Karen while I was consulting with him/her. We hired Dr. Addleman to review my medical records, interview me and write his medical expert report. An appointment was made for me to consult with him.

It was around this time that I joined the Cabot Toastmasters in order to become more experienced in public speaking. My mother, with the Newfoundland Teacher's Association, was an avid public speaker and my father was as well, during his over fifty years as a Mason. The first year I attended, I worked very hard to improve my skills and give speeches professionally. I would practice my speeches over and over again until I could speak for five minutes without notes. For my hard work, which was also helping regain my memory deficit, I won the Rookie of the Year award.

I was also grooming myself for the future when I would be speaking in the courtroom, in front of the judges, as I knew my medical malpractice case would never be settled out of court. However I did not realize that it would take fifty five days to present the evidence. Judge Halley informed me one day that I had more courtroom experience now than most young lawyers.

That year, I also joined the Positive Thinkers Club and enjoyed the monthly gatherings. One day, a new member, who knew me, spoke to me and said, "Why do you come to those meetings, you always appear so positive?" I laughed and said: "It is a positive thinkers club and I think positive." He said: "I am so negative and I came because I want to learn how to think positive."

I also arranged guest speakers and held monthly meetings for the Fibromyalgia Self help Group of Newfoundland and Labrador and joined the National Me/Fm Action Network, a national advocacy group for fm/cfs/mci.

I also felt I would like some media attention regarding my medical treatment and impending lawsuit, so I contacted the news media. A journalist from CBC interviewed me on three occasions, lasting over two hours each with video taping as well. I showed the journalist the medical records where the mistakes had occurred and I gave copies of the records freely. I though for sure CBC would help with my case, but unfortunately they stopped in their tracks and did not ever air any information about my case. The Telegram, on the other hand, were very interested and always wrote articles about the progression of my case in the courts.

Below is a copy of a letter that Dr. Karagianis sent to the journalist at CBC by facsimile.

December 11, 1998

Dear Ms ……

Re: Judith Day

My family and I are greatly upset by these allegations. This matter has already been carefully considered by the Newfoundland Medical Board and dismissed. In doing so, it is my understanding that the Medical Board agreed that my diagnosis and treatment of Ms. Day was appropriate.

I am gravely concerned that the reporting of this matter in public may affect my professional reputation, and therefore wish to emphasize that the claims brought by Ms. Day are allegations and <u>not</u> factual.

I feel sorry for Ms. Day's tribulations, however I believe my professional diagnosis and treatment of her medical condition was appropriate and correct.

Sincerely

James Karagianis, MD FRCP

Psychiatrist

On November 23, 1998, I had psycho neurological testing done at the Health Science Centre by Mr. Jim Woodrow. This testing was recommended by Dr. Rice due to cognitive losses that I had complained about for quite some time.

Exerpts from the report:
Mr. Jim Woodrow
Psychologist

Ms. Day did note substantive pain throughout the evaluation. Her pain was especially evident in manual dexterity task where she was unable to use her finger tips to place pegs on a pegboard.

I have lost my ability to do fine work with my fingertips like hand sewing, etc due to weakness, stiffness and pain.

We also noted a reluctance to answer unless she was absolutely positive. She had to be pushed in many tasks to make a response. Ms. Day notes specific concerns for poor short term memory as main result for the evaluation. No unusual signs or symptoms were noted.

Test results do indicate specific problems with memory. The WMSr General memory Index of 58 is below levels expected for an IQ estimate (FSIQ-96) Additional memory testing (word list, C) also noted memory problems (Complex Figure) also noted memory problems. The pattern of memory test performance suggested greater problems with verbal material.

Coupled with poor memory was exceptionally slow performance in manual dexterity task but not with mental speed tasks.

In conclusion, test results do indicate cognitive losses compatible with her complaints. She has difficulty with short term memory. It is difficult to gauge the extent of her problem, but I suspect the memory difficulties could interfere with academic pursuits as well as job performance. J. Woodrow M.Sc. Registered Psychologist

What shocked me the most about these test results was that during the medical malpractice lawsuit Mr. W. presented this report and then he stated: "I would question the extent of effort that Ms. Day had put into this examination." I believe the judge received the impression that I was faking during the testing and the results were skewed.

According to the defense's closing submissions, "on cross examination, "Dr." W. did acknowledge that depression or bipolar could explain a decreased memory function. And, more importantly, any of the symptoms Mr. W. noticed on his examinations could be attributable to bipolar disorder."

It was rather disturbing for me to hear these remarks, as Mr. W., a Health Science Centre employee, was not a medical doctor as he was being referred to. While he was called to court to present his test results, he was not there to give a medical opinion or diagnose symptoms. I realized then it is going to be very difficult to bring forth evidence to support my case.

To correct this false impression, I then asked a physiotherapist, who worked with Injury Management, to testify on my behalf regarding the 'real' effort I put into my therapy. She did some testing on the flexibility of my head, neck and shoulders that was accurately recorded by computer analysis. While presenting her evidence, she testified that I definitely did not fake any testing and put a considerable amount of effort into it and the rehabilitation. In fact, I had improved the flexibility of my head and neck from 11% in some areas to over 90% by the time the six month's therapy was completed.

To this day, I still perform those physiotherapeutic exercises, both in and out of water, and have kept the flexibility and mobility of my head and neck to within normal limits, mainly because it was stressed upon me by a neurologist that I could possibly have my driver's license revoked, if I did not improve the movement in my head and neck. I understood that as I knew I would be a menace when driving an automobile.

Throughout the winter of 1998-1999 I had been doing very well, but had bouts of flu like illnesses, in fact I was feeling like I was constantly having the flu. That would set me back with muscular pain and fatigue that would take me weeks to get my strength and energy back to carry on. I had another appointment to see Doucet as it had been almost six months since our last visit.

Consult Dr. Doucet
January 6, 1999 1636-1738
Bipolar Disorder

Generally energy is poor but variable.
Muscular fatigue and stiffness that is worse past few weeks due to flu.

Recognizes a pattern of her symptoms and discusses them at length, but attributes all of them to diagnosis of Fibromyalgia.

She is easily anxious and stressed. Not on any medication this past year.

She blames a lot of her symptoms, past and current, on misdiagnosis and treatment.

She denies anger, but will pursue legal action.

She recognizes a pattern of variable problems with memory, attention and concentration. She did have neurophysiology testing done.

She does not show any and denies thought disorder or pressured speech. She stays on track well.

She is trying to look after herself and has a careful lifestyle.

She denies any mood problems or current anxiety symptoms.

She was advised I disagreed with her views in that I felt Bipolar Diagnosis was correct and again reviewed why.

Plan: Will see PRN, whenever necessary

As I was not getting better physically and not well enough to attempt to go back to work full time again, I applied for Medical Disability Retirement, from my public Service Pension Plan and Canada Pension Plan.

I received this letter written February 19, 1999 from pension benefit specialist of the Government of Newfoundland and Labrador.

Our medical consultant has reviewed your assessments and is unable to make a determination on your disability.

Two assessments were submitted, one by your GP and one by your specialist. The lead diagnosis is totally different in both situations. From the specialist's perspective, the diagnosis is bipolar disorder and from the GP's perspective, the diagnosis is of fibromyalgia and, in fact, there is no mention of bipolar disorder in this application.

We require absolute clarification before we can proceed in that we have conflicting documentation. Basically we need prioritized lists of diagnosis.

If you require additional information please contact me.

I responded with this letter February 19, 1999

Dear Ms.

Further to our telephone conversation, another visit to my physician and an appointment made on March 11, 1999, with the psychiatrist, who is still insisting my diagnosis is bipolar disorder, I do not have a prioritized list of diagnoses.

As regarding a diagnosis of bipolar disorder, Dr. Amy Tong recommended, and Dr. Craig discontinued Lithium in August, 1996 due to toxicity. Except for taking Epival, recommended by Dr. Pryce Philips, a neurologist, for three months From October 1997 to January, 1998, when I was diagnosed with temporal lobe epileptiform activity, possibly resulting from Lithium toxicity, I have taken no other psychotropic drug, nor did I require any. Epival also exacerbated the fibromyalgia symptoms and caused my having two more admissions to hospital in January 1998.

It is very strange that if I do have bipolar disorder, I have lived all my life without drugs to control it and when I was prescribed drugs, I had severe toxic reactions to them (common symptoms of Fibromyalgia). Toxic reactions that exacerbated my symptoms of fibromyalgia to the point that I am unable to function well enough to pursue my career as a Nurse Educator or any full time job to support myself, which I have done all my life, living with undiagnosed Fibromyalgia.

Now I am totally disabled with Fibromyalgia. Symptoms I have coped with all my life until 1995, when I was prescribed drugs that interfered with the neurological networks of my brain, which are not understood and caused me brain damage, memory impairment, peripheral neuropathy, hyper mobility of my joints on my left side, stiffness, pain, weakness, vertigo, tinnitus, which are sometimes unbearable and left me with a reduced quality of life.

Because of my positivity and spirituality, I have learned to seek other ways, besides my career that I have lost, to fulfill my life and make it worthwhile, namely helping other people with Fibromyalgia learn to cope with their condition as I have.

I am disabled. What label or diagnosis is used in order for me to qualify for permanent disability is irrelevant. If the diagnosis is bipolar disorder, I cannot be forced to take drugs that are detrimental to my health. If the diagnosis is fibromyalgia, which I am being treated for at present, it does not make any difference, I am still disabled. But the negative stress that is caused by my having to fight for financial survival makes me sicker. I don't deserve that.

I had been gainfully employed for over thirty years until 1995, supported myself and my children, paid into disability pension insurance and federal and provincial pension plans and now I have to still fight for financial survival. I will work now towards changing this system.

If you need more information about my physical condition, you can consult Dr. Amy Tong and Dr. John Martin, both at the Health Care Corporation, both are specialists and both have diagnosed me with Fibromyalgia. Dr. Button, my family doctor whom I saw yesterday has completed the forms stating that I am totally disabled, after I had seen both of those medical specialists.

I felt now that Insult was being added to injury and I was being caused more grief by making me fight now for medical retirement. It would take another year before the government bureaucracy was in place for me to become medically retired.

When you are called to a sick man, be sure you know what is the matter. If you do not know, nature can do a great deal better than you can guess./ *Nicholas de Bellville*

Toastmasters Certificate of Achievement

Chapter 23

In Pursuit of Justice

I was not too far into my medical malpractice lawsuit when I realized the battle I was facing with the lawyers of The Canadian Medical Protective Association's (CMPA) and Health Care Corporation of St. John's (HCCSJ). I also discovered just how unfair and unjust the legal system is. Many lawyers kept telling me not to bother with this case, and advised me to give up due to the impossibility to win against the CMPA, and the possibility it would lead to bankruptcy, but I was determined to pursue justice and from the beginning, I had prepared for the worst.

The Canadian Medical Protective Association had become very wealthy and powerful since its inception in 1901. I learned that most of the malpractice insurance fees come from the taxpayers, through financial agreements between provincial governments and medical associations, when physicians are hired.

In 1998, I researched the financial statements and also learned that The CMPA paid out almost as much money in legal fees fighting malpractice suits as they did in compensation to injured victims, averaging at that time, around seventy five million dollars a year.

To discourage lawsuits, the mandate of the CMPA was to spend time wearing down the victim, the law firm representing the victim and force a settlement out of court with a minimum amount of money. The amount paid was usually barely enough to pay the legal fees incurred fighting these long and drawn out cases, usually lasting over five years.

The financial statement of the CMPA was reasonable proof that the defense's legal bills were as much as the payoffs, therefore the victims' legal bills are equally as high. The ordinary Canadian citizen could not win with this system in effect; therefore negligent physicians are being protected by their insurance company, and did not have to be accountable for their shoddy practices.

On February 16, 1999, I, with a few other concerned people, established and incorporated The Canadian Civil Protective Association, (CCPA), a non profit organization in order to help Canadians fight back against the Canadian Medical Protective Association, CMPA, a medical mutual defense organization to protect physicians against legal actions by their patients, based on allegations of malpractice or negligence, resulting in injury.

Although the actual figures remain unknown, it is estimated that 10,000 Canadians die annually because of medical errors and many thousands more are permanently injured and spend a lifetime suffering. There was such a culture of secrecy around these injuries and deaths, while little or nothing was being done to prevent them, and everything to increase them, while hospitals and nurses usually end up taking the blame, i.e. inaccurate charting, inadequate procedures, malfunctioning equipment, but rarely negligent physicians, until the injuries are so blatant that they cannot be overlooked.

Purpose of CCPA:

- Raise public awareness of the incidence of medical malpractice.

- Raise funds to assist victims of medical negligence resulting in harm by providing funding for legal fees, which will be reimbursed to the association if the case is settled favorably.

- Balance the scales of justice so that ordinary people have a chance to receive financial compensation when they have been injured by physicians.

The CCPA was neglected, as my compromised energy would not allow me to spread myself too thin. During that time, other groups across Canada were also emerging such as Canadian Health Coalition, Canadian Health Network. Voices of Health Care Concerns, (VoHCA) in Ontario and Dr. Susan B. McIver, Ph.D in entomology/ microbiology, with whom I corresponded many times, published her book <u>Medical Nightmares, The Human face of Errors</u>, and as a community coroner, listed many recommendations to ensure a better outcome for injured patients.

According to McIver, the examples in Ontario in 1998, total malpractice fees for specialists averaged $25,000 per specialty, depending on the risk factors of that specialty. In each category the physician paid less than $5,000 but the tax payer picked up the rest of the tab. I thought. Why did the insurance premiums need to be so high when at the time there were over four billion dollars in their account already and their budget was so low? This non-profit organization was definitely accumulating huge profits.

Also, according to McIver, even when there is an out of court financial settlement there should be some registry in place as a track record of physicians' past, their education, malpractice trials lost, dismissal from hospitals, disciplinary and remediation actions by the provincial College of Physicians and Surgeons, and any previously secretive settlements with former patients, among other relevant information. It is not good enough that these negligent physicians can just close their practice and set up in another province or state and keep on doing harm to patients.

If by chance a malpractice case is taken by a law firm on a contingency basis, the lawyer will usually give up and settle out of court before the payoff becomes less that his/her legal fees. Most of these battles last for years and at an average of $450 an hour lawyer's fees, these costs add up quickly. My malpractice lawsuit lasted from 1998-2008, costing me over $100,000, even without legal representation after the first year.

It is very difficult to obtain justice, because most patients or law firms do not want to risk losing any more money, fighting a lengthy battle in court. The CMPA have the best and most expensive lawyers in the country and medical experts supporting them, who are also members of this Canadian Medical Protective Association. Doctors band together against victims to avoid losing lawsuits, so then the doctors can keep their medical malpractice insurance premiums as low as possible, their assets high, and their reputations intact.

The Canadian Medical Protective Association, CMPA may believe what they were doing was legally correct, but it was ethically and morally wrong. I thought Canadians should become united, as the physicians have done to protect ourselves from this injustice.

I took on a second task at the same time; namely facilitating the Fibromyalgia Self help Group of Newfoundland and Labrador. I began to think that one would possibly be in conflict with the other anyway, so I concentrated my energies on working towards educating myself, other patients, physicians, health care workers and the general public regarding those invisible illnesses fm/cfs/mcs, so I put the CCPA on hold.

I was discouraged also when I was turned down by Canada Customs and Revenue Agency to allow it charitable status in order for me to issue tax receipts to the many donors who freely contributed to this important cause. I had received up to $500 for a single donation for CCPA, and had to explain to the donors that I could not issue receipts for taxation purposes as I was not granted charitable status.

According to CMPA's annual report, 2009, which I accessed on line, the figures paint a little brighter picture, but not good enough for the injured patient to get the compensation that he/she deserves. The awards and settlements are around $150,000,000 with the legal costs lower at $110,000,000, with expert consults being paid 12,000,000, and an overall asset of almost 5 billion dollars.

The average medical malpractice insurance fees are also reduced by 23% and there has been a dramatic overall decrease in the number of medical malpractice lawsuits filed from 1998-2008. This decrease was supposed to be due to better medical care resulting in fewer adverse events, increased awareness and understanding of patient safety, enhanced risk management procedures, and more effective and timely disclosure to patients and tort reform initiatives. I believe the latter is the most important to stave off medical malpractice lawsuits. In the beginning of my case, litigation was the last thing on my mind. All I wanted was the medical error to be corrected.

In March, 1999, I booked another appointment with Dr. Doucet, regarding my application for permanent disability as I needed his written medical expert report, so I visited his office in Churchill Square for eight minutes.

Consult Dr. Doucet
March 11, 1999 1752-1800
Bipolar Disorder

> **Feeling physically unwell**
> **EEG reported by Dr. Button**
> **She is awaking with pains in head. Right parietal temporal lobe.**
> **Feels increased weakness, worse left side.**
> **Energy poor, sets reasonable goals-easily drained.**
> **Sleep-fairly good.**
> **In spite of symptoms, she is exercising briefly daily.**
> **She has been organizing Fibromyalgia Support Group and another group Canadian Civil Protective Association.**
> **She does not feel she was ever manic or hypomanic, but blames all symptoms on drug reactions, stress and fibromyalgia.**

Another appointment was booked for two months, which I also kept.

Consult Dr. Doucet
May 27, 1999 1704-1730
Bipolar Disorder

> **She is coping fairly well.**
> **Denies mood symptoms.**
> **Complaining of fatiguing and muscular skeletal symptoms. Sleep is good.**
> **Bright, talkative, stays on track well**
> **Discussed concerns re diagnosis and controversial over this.**

During that spring, I was still consulting with my lawyer regarding the lawsuit and received many letters of correspondence between the lawyers as they were exchanging medical expert reports and information.

In June 1999, the defense sought an expert opinion of my medical treatment by Dr. Karagianis and Dr. Craig from Dr. M.D. Teehan, MD., FRCPC. And I received a copy. Dr. Teehan's opinion and summary were written from my medical records, which was based on inaccurate information being written and rewritten to support the diagnosis of bipolar disorder. The report was very damaging to my case as the errors and omissions on my medical files were not addressed.

After obtaining this expert opinion report, a Summary Trial was set for September 9th and 10th 1999, but the defense filed an application for a postponement of that trial and the judge ordered that the case be postponed indefinitely.

From the beginning of this lawsuit, I was educating myself how to do legal work as I was trying to cut down on the expenses. Each month since I hired a lawyer, I was averaging over $400 a month payment and I could not afford that, so I was doing as much work as I could possibly do myself. I believed in what I was doing as I thought that this lawsuit, whether I win or lose, was going to help me restore my cognitive ability and credibility, which it did. I had support from the courthouse staff members who were always courteous and helpful with directions and information when I needed clarification on those complicated legal matters.

I kept my appointment with Dr. Doucet On September 13, 1999 as I was directed to do by Blue Cross.

Dr. Doucet's Consult
September 13, 1999 1634-1728
Bipolar Disorder

Feels she is coping fairly well.
She has episodes of memory difficulty.
She has significant fluctuations in severity of the difficulties with memory.
Memory, concentration and information processing is decreased when physical energy is very poor.
Generally energy is poor and easily drained.
She is better at setting reasonable limits so she does not get too overtired.
She is still very preoccupied with issues of "being labeled with a mental illness."
Again, I reviewed this issue with her.
Plan: She will see me every two months and PRN.

My lawyer sought an opinion from Dr. Addleman in Fredericton, New Brunswick, as it would be convenient for me to travel there because my daughter Karen, was still living in Fredericton and I would have a place to stay.

I spent very little time in Dr. Addleman's office being interviewed, as there was a mix-up with my appointment time. When I received the medical expert report, I was definitely disillusioned with the medical establishment and their blatant disregard for the Hippocratic Oath. "First do no Harm."

Dr. Addleman Psychiatric medical expert report
September 18, 1999

Ms. Day is of the opinion that Lithium in particular is detrimental because of her fibromyalgia, and a note she received from Dr. D. J. Wallace, does state that, in his opinion Lithium can make fibromyalgia worse.

I was not aware of any evidence to that effect myself, and have been unable to find any research indicating that Lithium or any other psychoactive drugs edited by Bezchlibnyk- Butker and Jeffries, 7th revised addition, Hogrefe &Huber Publishers nor any of the monthly issues of the past several years of the Biological Therapies in Psychiatry newsletter.

Lithium is well known to cause a variety of neuromuscular side effects, including weakness and tremors, but these are usually transient and subside once the patient recovers from their illness and establishes a steady lithium blood level.

Toxic levels of Lithium can certainly cause significant neuromuscular problems, and some patients are more troubled by tremors and weakness than others, but I have had many patients suffering from fibromyalgia who had also taken mood stabilizers, including Lithium without any significant effect on their fibromyalgia. (I became neuro-toxic at normal blood levels, which those doctors did not know was possible.)

It is also clear from the record that although Ms. Day did take different medicines prior to these two hospital admissions there was never any evidence of a toxic, delirious state with disorientation or confusion that would lead to the conclusion that her symptoms were a result of these drugs.

It was clear from the first admission with Dr. Karagianis that the patient's concerns were acknowledged, and following her rapid improvement in hospital, her medicines were held, and she

promptly decompensated and again began displaying the same symptoms that had been prominent when she was admitted. They once again subsided when Lithium and Haldol were instituted.

Dr. Craig and his team were clearly knowledgeable that Prozac might have contributed to her manic symptoms when she was admitted to Health Science Center Hospital, and they promptly discontinued that drug.

She describes the recurrent problem she had from nasal congestion, nausea, motion sickness "a lot of colds and flu" and I think it is safe to assume that she used over the counter remedies for these problems intermittently without the precipitation of any abnormal emotional conditions. (This was an incorrect assumption. I consumed none after 1983, except the prescribed Trinalin that Dr. Peters, an allergist, had prescribed for me.)

In summary, In spite of the fact that Ms Day feels so negative about the care that she received, my reading of the notes indicates to me that both Dr. Karagianis and Dr. Craig recognized that she had significant problems with her health and did everything they could to help her recover.

From their interventions, it is clear to me that they both took her condition very seriously, and recognized that treatment was essential to prevent the illness from becoming recurrent and chronic and even more disabling.

Her psychiatrists both appear to recognize that she had many strengths and that there was a good chance that she would fully recover and have a good life, and this appears to have motivated them in their attempts to treat her.

Dr. Karagianis took his responsibility very seriously to the point that he admitted and treated the patient on an involuntary basis: Dr. Craig allowed the patient to control her treatment more than he wished.

Both physicians seem to have tried everything to enable the patient to persuade herself to stick with the treatment so that her mental condition would improve substantially enough that she would be able to manage her medical problems more effectively in the future.

Through self-care, I have recovered well enough to have a good quality of life, despite their damaging psychiatric treatment.

I believe Dr. Addleman did a grave injustice to me as none of the evidence that I had presented to him that was in my favor was mentioned by him and he made assumptions that were unfounded. It was very disheartening reading his medical expert report that was biased upon a subjective opinion. He paid no attention to the discussions we had during the interview and did his report on the charts only, discrediting my statements regarding my diagnosis of fibromyalgia and the side-effects of the drugs that I had been prescribed and not monitored properly, causing neurological damage.

Dr. Addleman charged me $2,500 for this medical report. Even the slowest reader could have read my medical files in six hours and $1000 could have been more than enough to pay for his work, which was useless, and caused my lawyer to give up on my case.

During the same time interval, my lawyer also wrote Dr. Frecker as he was the most objective and supportive of the entire psychiatrists who I had encountered over those past four years and since he had diagnosed me during the February, 1997 admission with "Adjustment Disorder" and then January 1998 as "questionable Bipolar disorder." despite what Dr. Teehan and Dr. Addleman have written on their medical expert reports that "all psychiatrists were in agreement with the diagnosis". All psychiatrists were not.

Dr. Frecker wrote the following to my lawyer, but sadly he died before my lawsuit came to trial.

This will acknowledge your correspondence of October 6, and November 19, 1999
I did supervise the care of Mrs. Day on two brief admissions at St. Clare's Hospital.
On these admissions she had been admitted through Emergency Department for further observation. My understanding was due to anxiety concerning her fibromyalgia and possibly some family tensions, she had become anxious.

Following admissions she settled down quickly and upon her own request was discharged from hospital for follow up with her regular psychiatrist.

I understand there has been a previous diagnosis on several occasions of bipolar illness, for which she has received treatment.

Opinion at the time of the admissions under my care was that this illness was in remission.

My opinion is that this illness was never there in the first place to go into remission.

My lawyer had advised me to drop the lawsuit as she thought that I would never have a chance to win it after she received that expert medical opinion from Dr. David Addleman. She felt she was ethically and morally obligated to withdraw her services. By now I had spent over $15,000 and had not progressed past the letter writing stage.

I wanted to carry on alone without a lawyer, as now it became more than a matter of winning or losing. It was an issue of physical survival, the possibility of ridding me of this label, and restoring my credibility. Now for the continuation of my own wellbeing, if the physicians do not want to admit their errors then I have to present my case to the courts, whatever the outcome. I want to have the evidence of my medical treatment on public record, if not, I will continue to be trapped within psychiatry and lose more control, credibility and dignity as the years pass.

I did not make this serious decision quickly or frivolously, I felt I had to continue with my quest for justice alone or I will forever be wondering what if? My thoughts are Christopher Reeve's thoughts "to be challenged is inevitable, but to be conquered is optional."

People's lives, like mine will continue to be destroyed if physicians do not take the time to correctly diagnose patients with such a serious illness as bipolar disorder.

I kept myself occupied with facilitating the Fibromyalgia Self-Help Group. Of course I obtained a great deal of satisfaction from this as I love helping and teaching others. Some days I had very little energy, but was satisfied if I just accomplished the task of getting out to the post office and mailing information packages to some of the people who would request them.

I kept my appointment with Dr. Doucet, November 12, 1999

Dr. Doucet's Consult
November 25, 1999 1607-1701

She is trying to stay active.

She is involved with local Fibromyalgia group. However she is setting reasonable limits for herself. She does better at pacing her energy and physical activity.

She is careful in diet and other lifestyle factors.

She feels her past diagnosis and treatment with label of bipolar disorder has been the most severe past and current stressor for her.

She blames it for losses in employment and professional standing.

She is still determined to address the question legally to "clear her name."

She denies depressed mood at any time.

Decreased energy, chronically with some marked fluctuations.

She is over talkative, but not pressured and thoughts show no abnormality in form or content.

Again I informed her that I felt her diagnosis was Bipolar Disorder and cautioned her regarding stressors.

Over that year, I planned and organized a Concert on Dec. 11, 1999, at the Arts and Culture Centre to raise public awareness about Fibromyalgia. Entertainers graciously donated their talents so I had exciting entertainment booked for the occasion, including the Mount Pearl Show Choir and Terry Lynn Eddy Band. This event consisted of informative sessions by Dr. Carl Misik, and Dr. Mike Carstensen, while Dr. Myrle Vokey was Master of Ceremonies. Furthermore, there were tables and booths arranged so providers of alternative medicine, and suppliers of health care products could advertise their goods and treatments that were available to treat chronic debilitating conditions like fm/cfs/mcs.

Most of the planning for this event was done on the computer or over the phone and it kept my mind occupied as I was putting together a public function for the first time. These activities were also helping me improve my memory. The concert was a success in my eyes and I accomplished what I had set out to do, that was to raise public awareness about this illness.

I had the memory testing done again February 18, 2000.

Mr. Jim Woodrow,

Psychologist.

Excerpts:

She denied any psychological distress in a mental health questionnaire. No pain symptoms were reported during the present evaluation.

The test results do note some improved cognitive skills. Performance in standardized tests improved. The IQ estimates improved from 96-105. The WMSr General Memory Index increased from 58 to 75. Timed manual dexterity tests improved from severely impaired to moderate to severe impairment range.

Unfortunately, I still do not have an adequate explanation for the problems. The current results would suggest that she is experiencing mild memory problems with motor slowness.

A week after my lawyer had notified the defense's lawyers that she was withdrawing from my case, they contacted me regarding initiating Discoveries and they would begin with me.

During my Discovery, which took place February 23, 2000, there was a representative from Health Science Centre management who knew me personally and she was literally crying while I was trying to bring forth my evidence. At that time, I was still having difficulty with my memory, especially when I was rushed, but with my note taking abilities, my organizational and filing skills, I managed to be interrogated for two hours by the defense and brought forward my claims, presenting the medical evidence, concerning the medical malpractice lawsuit.

On February 27, 2000, Ms Woodridge received a letter from The Medical Adviser, Dr. C. O'Shea handling my disability claim:

Dr. C. O'Shea

I have now received a short necessary report from Dr. Doucet, the attending psychiatrist on this case. I do feel now that I have sufficient evidence to justify permanent disability in this particular case and would suggest you do the same.

I have been medically retired since then and have never been investigated or questioned again regarding my ability to return to the workforce and be gainfully employed. I have kept physically and mentally active as an activist, a patient advocate, a writer, and an educator in health care issues, doing my very best in volunteer work to make my contributions to society.

With a diagnosis of being an unmedicated psychiatric patient with bipolar disorder, I knew it would be impossible to ever be gainfully employed again in the profession of nursing. What a loss of talent! What a shame!

The defense also discovered Dr. Eleanor Stein, my medical expert, Dr. Wayne Button, my family doctor, and one of the nurses from the Waterford Hospital. I believe the evidence that came from those discoveries should have warranted an out of court settlement for the mismanagement of my health.

During the spring of 2000, I also took an interest in politics as there was a federal bi-election coming and I thought it would be a good opportunity to become involved and see how I would make out being in the public eye and if I had the physical stamina to handle these situations as I still became tired so easily.

I felt this would possibly raise my profile and help restore my credibility that I felt had been completely destroyed. I had nothing left to lose and everything to gain, so I got involved with the up and coming Canadian Alliance Party and sought the nomination for the candidate to run in the federal bi-election in St. John's west. That idea was short lived when I disclosed my medical history and was gently eased out of the possibility to represent that party. From my experience, a psychiatric label is total ruination.

I felt I was compelled to consult with Dr. Doucet as it was a stipulation by Blue Cross, so I was not completely out of the psychiatric trap even after two more years of no psychiatric symptoms or taking no psychotropic drugs.

Dr. Doucet's Consult
May 18, 2000 1638-1732
Bipolar Disorder

> **Persistent poor energy.**
> **Muscular aches and pains with stiffness.**
> **Can walk around the lake, then stiff and poor energy to get out of car.**
> **Feels grip is weak at times.**
> **Difficulty up and down movements, and difficulty with balance.**
> **Denies problems with mood. Concentration and memory poor at times. Feels worsening in physical functioning.**
> **Again pressured and anxious to convey her point of view re illness and her legal case. Again, informed her of my opinion re her illness.**

I kept pacing myself, exercising, resting and following my self-prescribed care, but I was tiring easily and had pain and stiffness all the time. I began to realize that I needed to downsize my living quarters as I could not physically manage my two-story home as the many stairs were troubling for me. Also, without boarders and tenants, the cost of maintaining it was becoming financially impossible.

I found a tenant to rent the house and I downsized and moved into a two bedroom bungalow that was more manageable, with the option to purchase after a year. I managed to look after my daily chores, but I still had to hire people to do house cleaning and my heavy gardening.

That fall when the federal general election was called I announced on VOCM radio, where by now I had become a frequent caller regarding mostly the Fibromyalgia Self-Help Group of NL, that I was running as an independent candidate. I was mainly doing this to stress to the public the importance of Computerized Health Histories. I believed instead of spending money on a National Gun Registry, money would have been better utilized if it were used to set up a Computerized National Health History System. I believe if my own health history had not been scattered over five provinces, I may have been diagnosed correctly much sooner.

As was stipulated by Blue Cross, I had to consult with Dr. Doucet every six months.

Dr. Doucet's Consult
October 3, 2000 1617-1730
Bipolar Disorder

> **Persisting complaints of poor energy, pain and stiffness unchanged.**
> **Has been moved to two bedroom home.**
> **Less physically drained at home.**
> **Feels she is coping fairly well with her symptoms.**
> **Has continued and expanded her involvement with fibromyalgia Support group.**

She is offering her candidacy for up coming federal election Independent political party. Enthusiastic about goals.

?? overvalued but not grandiose or delusional.

She is anxious to defend herself from the diagnosis of Bipolar disorder and remains resistant.

Feels all her symptoms relate from chronic history of fibromyalgia.

I told her again that I don't dispute the diagnosis of fibromyalgia, but do support the diagnosis of depression (depression crossed out and Bipolar disorder written.

Caution about causing increased stress and risk to decompensate. She feels she is taking care of her needs and aware of limitations.

Plan: to see PRN.

I felt by reaching out to the public, I had definitely restored my credibility and had garnered enough votes during that federal election to entitle me to a refund of my deposit from Elections Canada. The experience was both rewarding and fulfilling for me and it certainly increased my profile.

Finally by the end of 2000, I notified Blue Cross that I had decided to get out of psychiatric care completely, and that I was not going to consult Dr. Doucet any more as I did not feel I required his services. My physical symptoms are real and I consider myself mentally stable and do not have mood swings. I have not seen a psychiatrist since then, and I have never been questioned regarding the validity of my long term disability.

I will continue to use most of my energy and resources in this battle to win damages for the impact that the mistaken diagnosis, confinement in a mental institution and incorrect drug therapy, has left on my life. I have gone through hell.

I kept working on my case alone, receiving informal advice from legal services, friends and relatives. In the beginning, I thought the courts were being fair and objective.

I worked through VOCM radio station open line programs to get health information out to public and reach the people who were newly diagnosed with fm/cfs/mcs, so that they could receive the information packages that I freely distributed to anybody who requested them. The Arthritis Society of Newfoundland and Labrador and Health and Community Services allowed me to photocopy at their expense the thousands of copies that I made of the information material that I distributed.

In the fall of 2000, when the Deputy Mayor White resigned, I thought I would like that role, so I ran in the municipal election, again attempting to raise my profile. I was honored when over 5,000 citizens of St. John's gave me their support. When Dorothy Wyatt former Mayor and Councilor at Large died during the election, and her seat had to be filled, I thought I may run again, but changed my mind, when I met Sandy Hickman, another contestant. I supported him all the way and he won the election. I then retired from politics to leave it to the young and energetic. At this time, I was neither.

The fall of 2000, I also hosted the 2nd annual Fibromyalgia awareness concert, which was the last one I attempted, as we did not break even with the costs and our annual operating budget was very little, we could not afford losses.

In 2002, I sold my house and instead of buying the bungalow I had rented, I moved into a condominium in a building, which I did not like. I felt I was in a box and wanted to be closer to the ground, so after a few months I sold the condo at a considerable profit, and bought a two bedroom bungalow, condominium where I lived until I moved to Fredericton, New Brunswick.

In 2002, from being affiliated with the National ME/FM Action Network through the Fibromyalgia Self Help Group of Newfoundland and Labrador, I discovered through their website, about Dr. Eleanor Stein, a psychiatrist who was specializing in FM/CFS. I contacted Dr. Stein by email and gave her a synopsis of my treatment by Dr. Karagianis and Dr. Craig and asked her if she would review my medical records. I told her that I believed I was misdiagnosed, maltreated and the drug treatment, sometimes forced, has resulted in my being permanently disabled from being gainfully employed as a nurse educator.

After exchanging emails and answering Dr. Stein's many questions, she agreed that she would review my hospital records, and if she found enough evidence to support my claims, she would write a report. From the beginning, Dr. Stein made it quite clear that I had to be totally honest with her as she was putting her career on the line. If she found me out with one untruth, she would back off immediately and not bother with the medical expert report.

Every piece of information that Dr. Stein wrote in her medical expert report, that supported my claim, could be backed up by medical evidence as you have read throughout the actual medical notes that I have rewritten in this book.

Below are excerpts from the summaries of Dr. Stein's Report which was over sixty pages, (a book in itself.) Dr. Stein wrote her report from June 18-July 12, 2002 that cost me over $8000. She charged me only $200.00 per hour and it was obvious that she put a tremendous amount of time extrapolating every bit of evidence that was required to win this medical malpractice lawsuit.

Examples below:

Written progress notes of Dr. Button, Ms. Day's GP from 1982 to present shows that throughout their 20 year relationship, Dr. Button only on one occasion (Feb/97 noted abnormalities in the mental status of Ms. Day.

Dr. Button wrote "Most of her symptoms fit more with fibromyalgia and over reacting to emotional situations."

At the time of each psychiatric admission, Ms. Day was taking some combination of prescription drugs to which she could have had a sensitivity reaction. Since discontinuing all drugs, she has had no recurrence of psychosis and no hospital admissions.

Treatment with Dr. Jain 1983 - 1994

In 1983 the patient (Ms. Judith Day) was admitted to the Health Sciences Centre (HSC) in St. John's under the care of Dr. Subash Jain. Dr. Jain summarized the admission and his outpatient follow-up of the patient over the following 11 years in a letter of March 26, 2001. Ms. Day's initial presentation was consistent with the differential diagnosis stated: "acute organic brain syndrome, mixed manic state or acute atypical psychosis (toxic confusional state secondary to medication and infection)". Ms. Day was treated briefly with antipsychotic medication and benzodiazepines. Ms. Day recovered quickly and was discharged within 14 days.

During the next 12 years, Ms. Day consulted Dr. Jain on an as needed basis. During this time she took no antipsychotic or mood stabilizing medication and Dr. Jain specifically notes in his letter of March 26, 2001 that "during out patient follow up with me, Ms. Day did not show any evidence of hypomanic or manic episodes. Ms. Day did not have any acute episodes of mixed or manic state or atypical psychoses as when she initially presented in 1983". "As there were no further episodes of psychoses during this time except mild and moderate depression she was not treated for bipolar disorder".

Given Ms. Day's presenting symptoms in 1983 the inclusion of bipolar disorder in the differential diagnosis was entirely understandable and appropriate. However her lack of symptoms, and her steady and excellent work record between 1983 and 1994 suggested that bipolar disorder was no longer the most likely cause of the 1983 admission. The conclusion reached collaboratively by Dr. Jain and Ms. Day that Ms. Day had had a psychotic drug reaction, was much more plausible.

Summary: The patient's treatment during the 1983 admission and the following 12 years was respectful and appropriate given the information available at the time. In retrospect, it is clear that Ms. Day suffered from Fibromyalgia during the time she was being seen by Dr. Jain and that much of her fatigue and stress actually was a result of the physical disorder and not due to depression. However the accepted definition for FM was not widely available until 1990, so Dr. Jain's attribution of the fatigue, sleep and stress problems to depression is understandable.

Consultation of Dr. Tom Smith, ENT (to Dr. Amy Tong) August 11, 1987

At this meeting, Ms. Day reported nasal obstructive symptoms, a history of feeling unwell and being tired all the time etc. Dr. Smith felt that the symptoms were likely allergic and referred Ms. Day to Dr. Peters, an allergist. This consult note established the presence of fatigue "all the time". Chronic fatigue is one of the cardinal symptoms of Fibromyalgia. Ms. Day was later diagnosed with this disorder.

Consultation of Dr. Sharon Peters, Allergist (to Dr. Wayne Button) August 13, 1987

This consultation established a long history of rhinitis, adverse drug reactions manifested by nausea and projectile vomiting

Consultation of Dr. Amy Tong, Internist (to Dr. S. Jain) Dec 20, 1993

This letter establishes current medications at that time: aspirin, Prozac ® and Trinalin ®. Dr. Tong notes history of adverse reactions to medications: codeine and alcohol both causing nausea. In review of systems, Dr. Tong notes "easy bruising, headache and shortness of breath and being tired all the time". Bruises were noted on the forearms and legs, abuse was denied as a cause of the bruises.

It is likely that if Ms. Day had been treated appropriately during the 1995 admissions as she was during the 1983 admission and medications had been withdrawn as soon as was safe to do so, that she would have been able to return to work and have been gainfully employed as she was from 1983 - 1995. However, due to the traumatic, harmful and protracted nature of her treatment from 1995 - 1997 Ms. Day is still recovering her ability to work.

To treat, as Dr. Karagianis suggests, only when diagnosis has been established results in substandard medical care. Kroenke et al published a well known paper in 1989 showing that only 16% of patients presenting to primary care with a variety of nonspecific symptoms such as: pain, fatigue, dizziness, headache, edema, back pain, dyspnea, insomnia, abdominal pain, numbness, impotence, weight loss, cough, and constipation were found to have a provable organic etiology (Kroenke & Mangelsdorff, 1989). Therefore treating on the basis of diagnosis would leave 84 % of patients with inadequate treatment. In these cases one must treat on a symptomatic basis.

Ms. Day's diagnosis was unclear at the time of admission and she had an underlying diagnosis (Fibromyalgia), which was not made by the attending staff. Therefore treating based on diagnosis resulted in harm, whereas listening to the patient and treating on a symptomatic basis, as was done in 1983 by Dr. Jain, may well have averted the severe consequences which resulted.

Due to lack of medical technology in general and lack of specific testing at the time of the injury (eg. failure to order an EEG at the time of lithium induced side effects were reported) it is not possible to prove that the involuntary drug treatment from 1995 - 1996 caused Ms. Day's current disabilities. There is only circumstantial evidence that Ms. Day's EEG and thyroid were normal prior to the treatment, and were abnormal sometime after treatment. Similarly it is not possible to prove beyond a doubt that Ms. Day's current disability was caused by the maltreatment during 1995 and 1996. However there was a significant change in the functioning of the patient during that time and the treatments she was given are known to have lasting side effects, especially in those who are highly sensitive as Ms. Day was and is.

The Mental Health Act was breached when Ms. Days appeal for release went unheard. If the facts of the case had been reviewed independently, then it would have been clear that there was insufficient evidence to certify and recertify Ms. Day.

Conclusions:

Summary of Events - Ms. Day voluntarily went to the Health Sciences Centre on May 15/1995 because she realized that she had become psychotic due to medications she had been taking for an upper respiratory tract infection. She was misdiagnosed with Bipolar Disorder even though the diagnostic criteria used at that time (DSM IV) for Bipolar Disorder were not fulfilled. The diagnosis was made by an intern at the HSC and was never properly investigated by Dr. Karagianis, who took over Ms. Day's care at the Waterford. The witnesses who knew Ms. Day, including her physicians were never interviewed. Information gathered from collaborative witnesses was misinterpreted. Ms. Day was detained illegally since she was a) a voluntary patient and b) not a danger to herself, others or property.

Abuse of power resulting in harm - Ms. Day was threatened with assaultive action (intramuscular drugs) if she did not comply with oral treatment and therefore complied, but suffered significant side effects from that treatment. This is a blatant abuse of power on the part of the physicians. The patient's repeated complaints of side effects were ignored.

The diagnosis and treatment during the Waterford admission influenced future care and Ms. Day remains labeled with the moniker of Bipolar Disorder to this day, even though since January 19, 1998, she decided to cease all medication and direct her own treatment.

The involuntary hospitalization and misdiagnosis of Bipolar Disorder caused Ms. Day pain and suffering through

- Alienation from her family,
- The psychological impact of having her basic right of freedom removed,
- The physical and cognitive effects of the involuntary medication.
- loss of her job and income as a direct result or misdiagnosis, stigma in the workplace induced by Dr. Craig's communication and drug side effects.

Misdiagnosis - Physicians like other human beings can and do make mistakes. Therefore the initial misdiagnosis of Bipolar Disorder rather than drug induced psychosis is not in itself negligent. It can be difficult to differentiate between psychotic conditions in the moment. However negligence occurred cumulatively over time when the patient's version of events, including her personal health history, was ignored and when she was treated against her will in a harmful manner over a 2 year period.

The diagnostic criteria for Fibromyalgia were published in 1990. By 1995, the diagnosis of Fibromyalgia should have been well known to all psychiatrists. Whereas Dr. Jain could be excused for misdiagnosing continued infection, pain and fatigue as signs of depression, this was no longer acceptable in 1995.

Never the less, both Drs. Karagianis and Craig would have avoided negligent behavior if they had listened to the patient, instead of insisting they knew best.

The dismissal of Fibromyalgia as a psychiatric or psychosomatic disorder is not acceptable. Deale & Wessely, conservative researchers from Oxford University in the UK have reported that "of patients with Chronic Fatigue Syndrome (a similar and overlapping disorder to Fibromyalgia) who had previously been given a psychiatric diagnosis, 68% had been misdiagnosed" (Deale & Wessely, 2000).

If this is true then a majority of the 500,000 persons in Canada with Fibromyalgia have likely been misdiagnosed. If Ms. Day's case is successful, health care for many of these may improve.

Medication Side Effects - Despite Ms. Day having been a highly functional nurse educator prior to the 1995 admission, Ms. Day's insight into her chronic sensitivity to prescription medications was ignored. Regardless of their personal beliefs about the concept of individual biochemical diversity and sensitivity to medication, the defendants should have noted the voiced complaints of the patient and the written observations of the trained nursing staff and acted accordingly by decreasing or stopping the offending medications.

I agree with the statement of Dr. Nurse who examined Ms. Day in 1997 and concluded, "I do not feel that any patient should be forced to take medication if they are not in agreement with the decision."

The issue of medication side effects is not insignificant. A National Academy of Sciences report in 'Scientific American' (May, 2000, page 16) stated that 'deaths in the USA due to medical error' may exceed 72,000 per annum. The article suggests that 'medical error' is the 8th commonest cause of death in America or the equivalent of five major air crashes every day.

In conclusion I find the case presented by Ms. Day to be strongly supported by the medical record. Several breaches of the Mental Health Act occurred. She was misdiagnosed and mistreated and suffered harm as a result. Her opinions were not listened to. She was not treated with respect. The health care system failed Ms. Day in that it did not follow the credo that has defined medicine since its inception "first do no harm".

_____   _____

Eleanor Stein MD FRCP(C)   **Date**

Calgary, Alberta Canada

When I presented Dr. Stein's medical report to the defense, the doctors' lawyers went looking for more heavyweight medical expert witnesses and found Dr. Joseph Berger and Dr. Keith Pearce where their main objective now was to disprove Dr. Stein's report.

Dr. Berger's report also consisted of sixty pages. These are some excerpt from his report.

My impression of reading the files, the testimonies of the psychiatrists being sued, and the independent assessment report, is that the psychiatrists behaved in a perfectly appropriate, professional, ethical manner in making their diagnosis and in offering treatment.

The fact that Ms. Day may challenge and may not believe their diagnoses does not affect in the least an independent assessment that the behavior of these two psychiatrists was appropriate, professional and ethical.

Their diagnostic conclusions and their treatment approaches were fully justified as judged by the material available to them at the time-according to the record that has been made available to me.

The notion that Ms. Day's symptoms may all be explained by the highly controversial label "fibromyalgia" has no good scientific basis to support it. The term fibromyalgia is not accepted within the overwhelming majority of current medical circles as referring to a medical illness, and its claimed manifestations have not previously been described as including such recurrent psychotic episodes.

Within psychiatric circles the term 'fibromyalgia' is taken as referring to a psychological condition expressed predominantly by the symptom of pain.

In my activity as an American Board Examiner, I have on some recent occasions asked my US colleagues – who are Deans of Medical schools, chairpersons of University Departments of Psychiatry, and Directors of Education in such departments- about the current views of their psychiatric and medical colleagues on those controversial labels such as "fibromyalgia" "chronic fatigue syndrome" "multiple chemical sensitivity."

My colleagues have told me that almost without exception, opinion is firm that these functional somatic syndromes as they are sometimes termed – are products of psychological conflict, not physical disease.

However, Ms. Day has managed to find one of the very few psychiatrists, in North America who does believe these are conditions of medical illness, and that psychiatrist, Dr. Stein has produced a very detailed assessment of the record in addition to Dr. Stein describing her own meetings with Ms. Day, and Dr. Stein is supporting Ms. Day's claims and allegations.

Dr. Berger writes five pages under the current scientific understanding of Fibromyalgia. This is his summary:

In spite of numerous claims, there is widespread agreement that no pathological lesions in the body have been consistently demonstrated.

There is widespread agreement that no abnormal biochemical or radiological abnormalities have been consistently demonstrated.

The large cross-over among all these patients (fm/cfs/mcs) has led a number of experts to conclude that a more useful descriptive categorization is to include them all under a heading "functional somatic syndromes.

These are truly wretched people, probably encouraged to get worse "or at least, not get better" by their choice of treatment centers. "In our book <u>Mediolegal Consequences of Trauma</u> (1992) Simon and I left out Fibromyalgia because the concept is not sustainable... we should not turn aspects of life, no matter how unpleasant, into disease: the physician should not become a co-conspirator." (Clinical Rheumatology. 2003.

I recognize that this is a complicated argument,(re the existence of Fibromyalgia), but it is necessary to understand why a patient or physician cannot just randomly claim "fibromyalgia is the cause of my symptoms" and expect to be taken seriously.

In Ms. Day's case the lack of logic becomes even more grotesque because there is a perfectly valid scientifically explanation for the symptoms Ms. Day has experienced over the years that have led to at least six known hospitalizations, and that is that she suffered from an intermittently recurring psychiatric illness known as Bipolar Disorder..

Dr. Berger Continues:

One of the most troubling aspects of contemporary medicine has been the tendency for patients to "diagnosis shop." Whereas in past times patients would accept what their physicians told them, or would at best ask for a second opinion.

Many of today's patients, armed with the latest lay synopsis from a magazine or newspaper article or internet reference, decide upon their own diagnosis and search for a physician who will agree with them.

Ms Day followed this path and in June 1998 she obtained a referral to Dr. John Martin, and told Dr. Martin that she believed that she had 'fibromyalgia all her life," even in childhood.

To base a claim that all Ms. Day's problems are a consequence of a so-called physical disease for which one can find no anatomical or pathological basis, appears to be far fetched indeed.

What is much more likely is that Ms. Day could not accept the diagnosis of a major psychiatric disorder. She has fought strenuously to avoid such a label. His notion that "fibromyalgia" could explain her symptoms is another "tactic" or maneuver in her fighting.

Ms. Day fails to appreciate that along with positing the wildest most esoteric theories to explain her breakdowns she also indicates behavior that is completely self contradictory and illogical.

These factors are very strong supports for the presence of a defective distorted perspective and sense of judgment that is more than just a history of temporary periods that are characteristic of psychotic episodes.

At the very least a diagnosis of Borderline Personality disorder would have been quite supportive based upon a number of features of history and behavior demonstrated during them, and the quite irrational conviction that a dubious controversial syndrome usually restricted to complaints of pain is responsible for the whole twenty year clinical picture. The diagnosis of a severe psychiatric disturbance is strongly supported.

Both Dr. Karagianis and Dr. Craig were fully justified in making a clinical diagnosis of Bipolar disorder. My impression is that Dr. Karagianis and Craig conducted themselves with a high level of professionalism in dealing with a difficult patient. I found the assessments to be justified by the data they had in front of them and they behaved in an honest and ethical manner, trying to come up with the best approaches and treatments. They requested the most appropriate investigations and consultations in a timely fashion.

The success of their approach can be seen in the relatively rapid recoveries that Ms. Day made from her state at the time of admission on those occasions.

However, Ms. Day's chronic disorder-though in abeyance in terms of severity for some years-eventually took its toll and in recent years she has been less able to function at work, and appears to have become obsessed with these notions of being badly treated, stigmatized, having a bizarre very controversial label as an 'underlying" condition, and pursuing litigation.

While this may have given her a 'cause" during the last few years, I don't think it has been of great help to her mental health. As noted, I also think that the ingratitude involved in this suit is a manifestation of her illness.

I found nothing in the material I read to support the allegations against Dr. Karagianis and Dr. Craig and neither did I find anything convincing at all in the analysis of Dr. Stein to support Ms. Day's allegations.

As noted, I found a number of Dr. Stein's allegations and expressions to be very unwise and injudicious. Her analysis of the material often appeared to focus on minor detail and miss the overall picture.

Ms Day has a serious recurrent mental illness, requiring a number of hospitalizations. She has been unable to accept the reality. Her psychological defense to that denial is to accuse others of misdiagnosis and mistreating her. From the objective perspective, there is no substance to her allegations.

Dr. Berger was not called as an expert witness.

It is not what a lawyer tells me I may do; but what humanity, reason, and justice tells me I ought to do. / Edmund Burke, 1775

Taken during one of my road tours throughout Atlantic Canada.

Chapter 24

The Trials and Tribulation

The defense initiated the Interlocutory Application (inter parties) (for summary trial) on April 9, 2002, to have the case dismissed for lack of evidence. I however, presented so much evidence that the judge did not dismiss the case. I was also advised by the judge to revise my Statement of Claim in order to support other accusations that came forward during the summary trial.

I finally brought my case to the state of readiness for trial and a date was set for November 12, 2003. I worked diligently for months preparing for the Pre-trial conference, where I thought all evidence would be weighed and a possible settlement would be in hand, as in my opinion, the evidence was overwhelming in my favor.

During the pre-trial conference, instead of discussing the trial as I had assumed, the discussion was centered on the fact that the physicians would not consider a settlement out of court, although the possibility of a mini-trial was addressed. I agreed to a mini-trial as I thought it would be more practical and cost effective, so it was set for the same date as the actual trial, November 12, 2003, five years after the case was filed.

In preparation for the mini-trial, the defense carried out Discoveries on physicians and nurses, and again the evidence that I had obtained for presentations supported my case, so I was feeling very confident about the outcome.

Four or five days were scheduled for the mini-trial, therefore I felt satisfied that the judge would weigh all the evidence before the courts and the outcome would be in my favor. I thought, perhaps naively, that in face of all of my evidence and Dr. Stein's expert testimony during Discovery he would 'advise' the defense to do the honorable thing and settle with me out of court.

Instead, I was surprised when I realized that the judge may have met with the defense lawyers before the mini- trial. I was of that opinion because Justice Barry had knowledge of insignificant details that he would not normally have been aware of unless he was told beforehand. For example, the term PWC, (person with chronic fatigue syndrome) as he referred to my medical expert, which according to him could reduce her credibility. This information was not written on any of the court documents.

The judge also referred to a "meeting" when he was discussing dates with the defense lawyers, for the following week to commence with the mini-trial. Also during the Pre-trial conference, the defense lawyers mentioned that they would like the mini-trial date set early so they could meet with the assigned judge.

Although I found out that this practice is legitimate, I thought to myself, how can a judge remain objective when he/she hears arguments and issues and receives information without all parties present? I was not present at this meeting!

During the mini-trial the judge seemed interested in one idea; to convince me to stop the legal proceedings (In his words) "bad things happen to good people and it is time to stop fighting and get on with your life."

I answered the judge by saying that "My fight was over in January 1998, when I was fighting for my life due to symptoms of drug toxicity from carelessly prescribed drugs and not being properly monitored for side effects". I stated, "Presently this is a challenge for justice so I can possibly receive the compensation that I truly deserve for the disabilities that I have to endure for the rest of my life due to this medical negligence."

The judge proceeded to inform me that "the courts" might consider my medical expert biased and therefore not be substantive in her testimony (notwithstanding her extensive credentials and reams of objective evidence that she had written in her report).

The judge then presented his own statements regarding some of the medical conditions in question, Fibromyalgia, (FMS), Chronic Fatigue

Syndrome, (CFS), and Multiple Chemical Sensitivities (MCS). This presentation appeared to me to illustrate both bias and a lack of medical knowledge on his part. I felt he was not qualified to make any remarks on those subjects.

During this mini-trial, the report of a defense medical witness, Dr. Keith Pearce was accepted as credible by the judge, yet I presented evidence that this witness had been formally reprimanded in the past by the Courts in Alberta for the validity of his testimony and biased opinion, ignoring scientific research, regarding those medically diagnosed physical conditions of fm/cfs/mcs. I presented to the judge the actual court cases from Alberta. These cases are listed below.

- In Baillie v. Crown Life Insurance Company (1998), 59 Alta. L.R. (3d) 45 (Q.B.) at para 33 et seq., Clarke J. took issue with Dr. Pearce's view that chronic fatigue syndrome "does not exist".

- In Phillips & Phillips v. Rost (1996), 40 Alta. L.R. (3d) 246 (Q.B.) at para. 152, Moreau J. rejected Dr. Pearce's opinion that the plaintiff was not "forthright" and Dr. Pearce's attribution of the plaintiff's "on-going problems to hysteria, a desire for secondary gain, and malingering".

- In Reynolds v. Pohynayko et al. (1997), 202 A.R. 1 (Q.B.) at para. 53, Bielby J. wrote, "Plaintiff's counsel launched an effective attack on the objectivity of Dr. Pearce, a psychiatrist retained by the Defense to test and interview [the plaintiff]."

- In Stevens v. Okrainec (1997), 210 A.R. 161 (Q.B.) at para. 254, Nash J. stated, "I do not accept the opinion of Dr. Pearce that [the plaintiff] has pain of a non-organic origin that was caused by her personality problem.

To make matters worse, there is no audio taped record kept or documentation of mini-trials. The entire mini-trial lasted two days instead of the scheduled four or five.

How can the legal process be open and transparent if private "off the record" meetings can be held before mini-trial proceedings, and then judges can make statements and opinions that cannot be held for scrutiny as no formal record of them exists? Is this justice?

I personally do not believe that any conversations between judges and lawyers about impending trials should take place without all parties being present, and without being audio or video recorded as judges, being only human, may have a tendency to be swayed in the wrong direction.

Judges may give an opinion that could be very detrimental to one side or another if it is made without studying every detail of the evidence presented to the court, or listening to both sides of "all" the evidence, which was not done in my case during the mini-trial.

Even though after a mini-trial, the judge's opinion is not a formal decision, it still bears down heavily on the consequences of the case, and in turn, I believe in my case, it obstructed justice instead of upholding it.

I wrote a letter to Justice Barry, January 24, 2004, pointing out the facts as I have written above and I received a response February 3, 2004.

First Justice Barry wishes me to point out that your allegation he met with defense lawyers prior to the mini-trial is false and apparently based upon your mistaken interpretation of facts or your irresponsible assumptions.

Second, Justice Barry notes your continuing confusion about the purpose of a mini-trial and suggests that you consider Rule 39:05 which requires the judge at a mini-trial to give a 'non binding opinion on the probable outcome of a trial."

The opinion must deal with expert opinions as well as other relevant matters and the opinion must be communicated to all parties participating in the mini-trial.

You will receive no further communication from Justice Barry, concerning the mini-trial.

I then had to apply for another trial date, for which I was in no hurry. I felt, the more time that could pass without my experiencing any mental health issues, this alone would help prove my case.

In the fall of 2003, with my daughter, Karen's encouragement, I decided to sell my home and move to Fredericton as she and Tim were building their new home and I wanted to buy back the home that I had purchased in 1994 for my retirement living.

By this time, I had obtained another lawyer Bob Buckingham, gave him a retainer fee of $1000, and my many boxes of court files. I moved to Fredericton around the end of March, 2004. I left behind my many wonderful friends and loving family, including my son Brian and his immediate family, Sabrina, their son Roman and Sabrina's large family of siblings, parents, grandparent, uncles, aunts, cousins and friends from whom I have always been welcomed in their homes with an abundance of affection, support and laughter.

I settled in my new home in Fredericton, and started immediately to create my gardens as it was a gardener's paradise here in Fredericton with the warm and sunny spring and so much beauty everywhere. The spring and summer flew as I spent every waking hour possible in my garden, rain or shine. My summer could not have been more therapeutic and uplifting.

In July 2004, Dr. Stein wanted me to have a consult with Dr. Flor Henry who was doing scientific research at the Alberta Hospital, Edmonton. I made an appointment, and travelled to Edmonton.

I had a very lengthy interview with Dr. Flor-Henry and he reminded me of Dr. Jain. He was very attentive and I gave him the details of my past few years and how the diagnosis and treatment for bipolar disorder had made me so ill. The testing went on for a few more days. Dr. Flor-Henry did scientific investigations and I was included in his research. Below is his report of the results:

The Quick Diagnostic Interview Schedule (DSM-11R) a systematic computerized mental state examination, only one psychiatric disorder was elicited: somatization (because of somatic symptoms, fatigue and pain.

The basic personality Inventory (BPI) is a computerized self administered test that measures twelve personality dimensions, and where the subject is compared to general population of North America women. All the dimensions were the same as all North American adult females.

On the Multidimensional aptitude battery (MAB) with a verbal IQ of 102, Performance IQ of 97, the subject falls in the average range of general intelligence.

Electodermal psychophysiology in this dextral subject was abnormal because of slow habituation this suggesting relative right brain preponderance. All the other psychophysiological parameters, Somatosensory evoked potentials were normal as were visual evoked potentials and auditory stem potentials. The 40 Hz Galambos procedure was also normal.

Centroid analysis indicates that the EEG is statistically abnormal in all frequency bands in the Eyes open and Eyes closed conditions. During the Word Finding task it is abnormal in the delta and beta bands. (2-3Hz, 14-20Hz, 21-50 Hz) The same is true during the Dot localization Task. Inspection of the EEG maps, looking in each of the four conditions the one that is statistically most abnormal (interestingly this turns out to be the 14-20 Hz band in all cases), there is increased power posteriorly L>R. The pattern is different for the delta band (2-3 Hz) in the Eyes Closed condition where there is again increased power in the posterior regions but also over the whole of the right hemisphere. The side view left and right shows dramatically the increased power in the right hemisphere in the slow frequency. EEG coherences are generally increased in the left temporo-parietal region.

On a discriminant function analysis when the subject is compared to a group of unmedicated women with depression the EEG falls fully in the depression group in all four conditions and in all frequency bands. The same is true when she is compared to a group of women with mania (except for delta and theta during the Word finding task). When compared to a group of women with chronic fatigue syndrome, she again falls fully into the fatigue group in the alpha and beta frequencies. This is also seen when she is compared to a group of women with fibromyalgia.

In a different analysis the spatial distribution of the voltages of the EEG (across 1-30 Hz frequencies) is compared on 8 channels to a bank of unmedicated schizophrenics, manics, depressives and normals. Two discriminant functions are applied: quadratic and linear. Here the configuration is clearly manic-depressive.

A cerebral blood flow study was done at the U of A (SPECT) on July 16, 2004 and showed "an abnormal unusual focus of activity at the right globus pallidus".

The diagnostic impression is of manic depressive illness and chronic fatigue syndrome/fibromyalgia. From the information available to me (reports of the patient, and medical and hospital records) there is a strong impression that her side effects and toxicity exhibited by the patient to lithium and neuroleptics (gastrointestinal, CNS, thyroid) were largely dismissed or minimized throughout her many years of treatment. The temporal EEG abnormality was clearly induced by lithium as lithium toxicity can induce convulsions.

Whether this scientific experimentation and examination by EEG done by Dr. Flor Henry and the recordings are used to diagnose manic depressive illness or not, or if the abnormalities on my testing were due to the drug toxicity alone, one can only guess, but I still have no symptoms of mood swings indicative of manic depressive illness and I had been off all psychotropic medications by then for over six years.

The rest of the summer of 2004, I chose to forget about the impending trial as I had left it in the hands of Bob Buckingham until I received an email from the defense, the second week of December, 2004, with an attachment from the court notifying me that the trial was set for January 11, 2005 and I had to advise the court immediately when I would be available for a Pre-Trial Conference some time between December 13 and December 20, 2004.

Apparently, I had not received the notification of this trial as the court had no record of my mailing address in Fredericton and somehow was not aware, even though I had informed them of both, that I had retained a lawyer. He was not notified either.

I phoned the court immediately and booked the pre-trial conference for Monday, December 20. I then packed into my car my medical files, my laptop and printer and whatever belongings that I would need for the next month or so, and began the long drive to Newfoundland just before winter had set in.

When I reached St. John's I met with my lawyer who had not had a chance to work on my files, so with such short notice, he could not even accommodate me for the Pre-Trial Conference, let alone a trial that was commencing January 11, 2005.

I then decided that I would proceed alone with his willingness to support me and be there with any advice that I would need. He also provided a photo copy machine with supplies that I could use whenever I needed copies. He also appeared in court when the judge requested that I needed a lawyer to interpret the law, even though I had become pretty good at court work by reading books from the law library at University of New Brunswick, and asking the law clerks questions. The lawyers who knew me, and knew what I had gone through with psychiatry, kept saying to me "if anyone can win this case regarding a mental illness issue, you have the best chance of doing it."

I will be forever grateful to Bob Buckingham for all he did for me, even giving me back the $1000 retainer fee just before Christmas 2005 that provided me with the money to buy Christmas gifts that year.

I believe, deep down, we all suspected that the case would be impossible to win due to my mental illness diagnosis. If anybody could convince the courts that I was indeed not mentally ill, and an injustice was done, it definitely would have to be me alone.

Dr. Stein mentioned that fact in her testimony: "It would be highly unlikely for a person who has untreated bipolar disorder to endure the stress of a trial for that period of time, without relapsing." When the trial began, I was representing myself, but I felt very confident in what I was doing and understood the challenge that lay before me.

The first day in the courtroom, the defense made an application to Judge Raymond Halley who presided over the trial to have Dr. Stein's report thrown out because; in their opinion she used pejorative language and was biased. I asked the judge if he would consent to the report being rewritten by Dr. Stein with the identified pejorative language removed. The judge did not consent to it, and threw out her medical expert report, but did allow her to testify and be cross examined. Dr. Stein was on the witness stand for five days.

After a few days into my personal testimony, I broke down in tears when I mentioned my mother's death. Judge Halley asked me if I had somebody who could come and be with me as he noticed I was all alone there and I said "yes, but I chose not to put my friends and family through this grief again by their relieving the trauma through my testimony." After that my friends came and sat with me, but found it very stressful, knowing me so well, and hearing the controversial remarks against me.

I spent thirty three days from January 11- May 13, 2005 bringing forth other expert witnesses to support my case including 13 specialists, 9 family members and friends, plus I spent six days giving personal testimony and being cross examined, to back up my claims.

After I presented my evidence, the defense made an application for a Non-Suit Ruling, in which the court proceedings can be dismissed on the grounds "that upon facts and the law, no case had been made out." but Judge Halley decided that I had presented enough evidence to prove my case.

The defense also asserted that the Statute of Limitations regarding the time to file my lawsuit (1995-1998) had exceeded, but due to the fact I had been so physically ill for most of that time and did not become aware of the mistakes until 1998, after I had read my medical files, it was not valid. The judge ruled in my favor on both counts and allowed the case to go forward, which did not happen for another six months. I believe that period of time was too long for the defense's evidence to be presented after the judge heard mine.

From November 21- December 16, 2005, we spent another fifteen days in the courtroom with the defenses' witnesses including among others, the medical experts Dr. Teehan and Dr. Pearce and their reports

Dr. M. D. Teehan's Expert Report
Written June 11, 1999, testimony November, 2005

Having surveyed the various hospital charts, the detailed clinical recordings by the physicians and other clinicians involved in the case of this patient, I have no doubt that she has quite a classic Bipolar Mood Disorder.

Not only are the clinical features described typical of the episodes, mania, depression and mixed episodes, but the pattern of her illness is also classical, with recurrent episodes, apparently some of them provoked by stressful events in her life.

Unfortunately, the patient has rejected the diagnostic conclusions of several different psychiatrists, who have all agreed that the diagnosis is correct.

As mentioned earlier in this report, there has been consideration of other potential causes of the changes in behavior and thinking that she has experienced.

The patient has proposed at various times that her illness was caused by antihistamine medications, and laterally Fibromyalgia.

To accept that proposition that her illness was caused by antihistamine medications, one would have to accept that several different compounds produced identical clinical features to mania, identical episodes separated by 12-14 years, and that as a separate issue she developed periods of depression.

Ranked against this possibility is the opinion of a number of unrelated specialists in Psychiatric Medicine, who separately and without collusion identified the presence of features of Bipolar Mood Disorder, on separate occasions, and who showed evidence in the clinical notes of their work that they carefully evaluated her history and mental state and relevant investigations in making that diagnosis.

In reflecting on the question of whether the physicians cited in the case have exercised due diligence and skill in the carrying out of their care of this patient, I found myself reversing the question: In other words I asked if the physicians had ignored the very obvious signs of a common and well recognized mental disorder, and instead had assigned diagnosis of antihistamine induced psychosis, would they have met the standard of accepted medical care? I believe there would be a real case to answer for practicing below the standard, if they had weighed the evidence in the direction of accepting that most unlikely proposition.

Subsequently, the patient has maintained that her episodes of manic behavior, with psychotic features requiring admission to hospital are the result of a recent diagnosis of fibromyalgia.

Once again to accept this proposition, one would have to set aside or ignore the pattern of cyclical episodes of mood disturbance of a classic kind, which is the recognized hallmark of Bipolar Mood Disorder, and instead choose a diagnosis of a predominantly physical illness to explain these features.

Again, had the physicians in question accepted the latter explanation of the symptoms, I think there would be a legitimate case for examining whether the standard of care had been breached? Had either of the physicians failed to offer the patient continuing, ongoing mood stabilizing medications, I believe that this would have been substandard care. With the weight of evidence put before them of Bipolar Disorder, to fail to address the risk of recurrence would amount to depriving the patient of needed standard medical care.

Furthermore, the management of the acute episodes of illness was essentially driven by the symptom picture present. Whatever the cause of those symptoms, it would have been necessary to treat the persistent problems of insomnia, over activity, delusional ideas and loss of contact with reality. The treatments applied are standard and no unusual choices of medication or combinations were selected.

From my examination of the evidence that you have supplied to me in this case, there is an overwhelming impression that both of the named physicians carried out their diagnostic formulations in very conventional ways, and arrived at reasonable conclusions and treatment plans.

The other physicians whose clinical notes appear in a documentation are in clear support of those diagnostic impressions, and in no part of the documentation from any clinical staff who have been part of the management of this case, is there any suggestion that there was fundamental disagreement with the diagnosis or treatment plan.

I would therefore have to say that I find no evidence that either Dr. Karagianis or Dr. Craig deviated from the expected standard of care for psychiatrists managing a case of this kind, and that they have followed well accepted clinical paths for this challenging case.

Upon examination, Dr. Teehan testified that it was unusual that I was transferred to Waterford Hospital, May 15, 1995 by taxi, with a nursing assistant, if I were really in a manic phase of bipolar disorder.

These are Excerpts from Dr. Pearce's medical expert report:

The procedures followed by the above professionals were entirely in keeping with the highest standards of psychiatric care. The carefully recorded clinical observations reflect throughout with no room for doubt, bipolar disorder.

At no time did either of them prescribe treatment, particularly drug therapy, of a kind and in amounts that were dangerous and detrimental to the health of the patient. The drugs primarily used- haloperidol and lithium carbonate- are well known to be associated with certain side effects.

In the case of haloperidol, the side effects are reflected in a change in the tone of voluntary muscle, so that they tend to be rather rigid, or perhaps most commonly, develop a type of tremor.

In the case of both psychiatrists, when they prescribed haloperidol in a significant amounts they invariably added a prophylactic dose of an antiparkinson drug to combat these effects.

In my review of the records in the case of both psychiatrists, I was unable to find any record of the presence of these side effects being noted, the routine prescribing of the antiparkinson medication seemed to have acted in the desired way to prevent the emergence of any tremor or stiffness.

In the case of Lithium Carbonate being used as a mood stabilizer, Lithium will only produce dangerous side effects when it exceeds a certain blood level. A blood level lower than the therapeutic level will have no benefit, but does expose the patient to the same risk of developing renal and thyroid problems. These if recognized promptly, are easily remediable, as occurred under Dr. Craig's care.

Unfortunately the lack of insight, which was an element in the diagnosis of Ms. Judith Day, prevented her from being able to cooperate fully; however at no time was there any recording which suggested she was experiencing toxic side effects from lithium.

When the possibility of lithium side effects became evident to Dr. Craig, his management of Ms. Days' medication was correct, prompt and ethical.

Under these circumstances, the maximum benefit will not be achieved. The results will always be less than optimal, despite the best possible standards of care by psychiatrists.

I am not aware of any cold or flu medications that has been recorded to precipitate an episode of mania in another wise normal individual, but we do have episodes of mania occurring when there was no evidence of such medication being used and causing hypomanic or manic episode.

What is very interesting about fibromyalgia is that the patients who suffer with what previously had been diagnosed as that condition, are in fact hypersensitive to pain- a finding that is universal in depression. They also seem to have a personality which, in times past, might have been described as hysterical, but which today we can say are characterized by hypochondrias or somatization.

Hypochondrias is, of course, worrying about a disease that has no psychical basis in reality and is expressed in what used to be called psychosomatic symptoms, which are devoid of any physical pathological basis.

We now recognize that a number of disorders can be fitted under a general umbrella, for instance irritable bowel syndrome, certain kinds of migrainous headaches, and chronic fatigue syndrome, as well as a number of sexual disorders.

Because of widespread prejudice against mental illness- most of us would be reluctant to acknowledge they had a mental disease or disorder, but certain types of personality are extraordinarily not reluctant to do so, so that it is apparently easier for them to accept some pseudo-physical diagnosis as a cause of their difficulties, rather than a clearly identified and eminently treatable, brain disorder, such as bipolar disorder.

Conclusion:

The records maintained by Dr. Karagianis and Dr. Craig and varied hospital staff, which I have reviewed in detail, in my opinion reflect the highest standards of psychiatric care, including diagnosis, the up to date treatment for bipolar disorder diagnosis they made, and a level of human concern for the patient's welfare which was of the highest ethical standard.

Since Dr. Karagianis and Craig treated Ms. Day, several other psychiatrists have attempted to treat her with a similar lack of lasting success.

Other questionable comments were taken from this report and written in the Judges' Decision In the next chapter. Dr. Pearce was well experienced in presenting expert opinions to the detriment of many patients in Alberta. He was absolutely ruthless, even approaching me personally when I was alone in the courtroom informing me that everybody was laughing at me. I responded "I don't think so Sir. This is not a laughing matter." I turned around and walked away from him without another word being spoken.

Both the defense's and my medical expert's opinions could not have been more opposite, completely bipolar (opposite poles). All the other evidence that I had presented in my favor was not accepted by the judge, not even mentioned in his decision.

A judge's decision, when it comes to expert witnesses, is a matter of who is more convincing, as a judge may have little or no knowledge regarding medicine. The case boiled down to the judge accepting the defenses' expert opinions of Dr. Teehan and Dr. Pearce over Dr. Stein's. There was absolutely no weight given to my testimony or the testimonies of all the other witnesses, so the judges' decision to accept the defense's experts went in their favor.

Oliver Wendell Holmes stated once "This is a court of Law, young man, not a court of Justice." Those words are so true.

No man suffers injustice without learning vaguely but surely what Justice is. /Isac Rosenfeld

The supreme Court of Newfoundland and Labrador

Chapter 25

Appeals and Closure

I filed a Notice of an Appeal September 13, 2006 on the following grounds.

- **The judge erred in fact and law in his decision that I failed to prove that the Defendants were negligent in the care and treatment.**

- **The judge erred in believing the main issue in the trial was whether the diagnosis of bipolar disorder was appropriate or not; and believing that if the diagnosis was appropriate, then the treatment, however cruel and harmful, was also appropriate.**

- **The judge erred in declaring the certificates of involuntary confinement signed by Dr. Karagianis, Dr. Pratt and Dr. Coovadia were valid with contrary evidence presented, and whether or not the certificates should have remained in effect for five weeks.**

- **The Judge wrote inaccuracies and misunderstandings in his decision to the extent that I believe he lacked the understanding to formulate the correct decision**

Bob Buckingham advised me to consult with Ches Crosbie Law Firm for his opinion on the grounds for an appeal. Mr. Crosbie advised me to drop the lawsuit against the Health Care Corporation and that the only ground that he could see was that the judge erred in law, and in fact, in that he failed to consider whether the Defendants negligently omitted to include medication effects in their working or final diagnosis and treatments.

Mr. Crosbie also consulted Dr. Charles Hutton the Chief Forensic Pathologist at the Health Science Centre, who I had known personally and had worked with him on medical-legal matters in the Operating Room since the eighties. Dr. Hutton had also signed my nomination papers when I entered municipal politics.

After we signed the papers, Dr. Hutton and I went out for lunch. He said "Judi what are you going to say when people ask you about being a patient at the Waterford Hospital? I replied "I will refer them to you."

I was rather disappointed with the remarks Dr. Hutton had put on a hand written letter to "Ches," dated November 23, 2006.

I expressed my concerns to Mr. Crosbie regarding the below documentation to Mr. Crosbie. If Dr. Hutton, who had not read my medical history, had gleaned this information from reading Judge Halley's Decision, the judge had definitely misapprehended vital evidence. I also pointed out the <u>errors</u> identified on Dr. Hutton's letter to Mr. Crosbie below.

I read Judge Halley's Decision twice. Judi had seven major involuntary admissions from January 15, 1983 to January 9, 1998. (This is untrue).

Each admission was for the "classic" signs and symptoms of manic phase of bipolar disorder. (There is no evidence of this on my file. In fact, according to Dr. Eleanor Stein, I did not meet the criteria at all)

In that time, nine psychiatrists from three different institutions (HSC, St. Clare's and Waterford all concur that Judi has bipolar disorder.

(A snowball effect had resulted and nobody corrected the error. Other diagnoses were written as "discharge diagnosis" eg. Stress induced anxiety, adjustment disorder and the last diagnosis in January 1998 read questionable bipolar disorder.)

Two expert psychiatrists for the defense with heavy CVs reviewed the mountains of medical records and also concur that bipolar disorder was the correct diagnosis and the involuntary confinements justified and met the standards outlined in the Medical Act. They both agreed the management and medications were appropriate. (You be the judge)

Judi's only expert was Dr. Stein, who is a psychiatrist, but mainly child and adolescent practice. She now confines her practice to fibromyalgia and chronic fatigue syndrome. She has only one month experience in adult psychiatry. The judge considered her evidence more as an advocate for Judi. No contest in the Battle of Experts. No negligence, no case.

Dr. John Martin the Rheumatologist confirmed the diagnosis of fibromyalgia in Judi. So Judi is a patient with bipolar disorder and fibromyalgia. I agree.

Unfortunately Judi will not accept the fact that she has a bipolar disorder and when not compliant with medication, she gets worse and when she is compliant she gets better. This is documented in the medical record. (The opposite is documented in the medical record.)

I don't see what there is to appeal. I can't understand why there was a 55 day trial. Whoever advised Judi to go with this case must be incompetent or out of money.

Upon Mr. Crosbie's advice, I dropped the lawsuit against the Health Care Corporation of St. John's and filed an amended Notice of Appeal in the Supreme Court of NL, Court of Appeal, dated November 23, 2006.

- **The Learned judge erred in law and in fact in that he failed to consider whether the Defendants negligently omitted to include medication effects in their working diagnosis**

- **The learned trial judge erred in law in that he failed to make a provisional assessment of damages**

- **The learned trial judge erred in law in that he failed to provide reasons for decision sufficient to permit meaningful appellate review.**

I produced a seventy page factum of the inefficiencies, January 8, 2008.

The judges ruled in favor of the defendants. I was also ordered to pay the entire costs that were incurred by the Canadian Medical Protective Association. Below is a copy of the appellant judges' decision,

Docket: 06/99 & 06/101
Citation: 2008 NLCA 32

IN THE SUPREME COURT OF NEWFOUNDLAND AND LABRADOR COURT OF APPEAL

BETWEEN:

JUDITH DAY APPELLANT/
RESPONDENT BY CROSS APPEAL

AND:

JAMES KARAGIANIS FIRST RESPONDENT/

FIRST APPELLANT BY CROSS APPEAL

AND:

DAVID CRAIG — SECOND RESPONDENT/ SECOND APPELLANT BY CROSS APPEAL

AND:

THE HEALTH CARE CORPORATION OF ST. JOHN'S [DISCONTINUED] — THIRD RESPONDENT

Coram: Wells, C.J.N.L., Welsh and Rowe, JJ.A.

Court Appealed From: Supreme Court of Newfoundland and Labrador
Trial Division 1998014087

Appeal Heard: May 21, 2008
Judgment Rendered: June 24, 2008

Reasons for Judgment by Rowe, J.A.
Concurred in by Wells, C.J.N.L. and Welsh, J.A.

The Appellant/ Respondent by Cross Appeal: Self Represented
Counsel for the Respondents/Appellants by Cross Appeal: Jane Hennebury

Rowe, J.A.:

INTRODUCTION

[1] Judith Day is appealing a Trial Division decision dismissing her claim for professional negligence against two psychiatrists who treated her.

FACTS

[2] The essence of Ms. Day's claim is that various psychiatrists, notably Dr. Karagianis and Dr. Craig, misdiagnosed her condition. Their diagnosis was that Ms. Day has bipolar disease, a mental disorder; they treated her with medications in line with their diagnosis.

[3] Ms. Day's position is that she does not have bipolar disease, but rather had a series of toxic drug reactions (notably to the medications prescribed to her for bipolar disorder) and that she also has fibromyalgia and chronic fatigue syndrome.

[4] Ms. Day's position is that the medications prescribed by Dr. Karagianis and Dr. Craig caused her serious harm (physically, emotionally, in her career and in her personal life), for which she has sought general and specific damages in total just over $2 million.

[5] The facts were summarized by the Trial Judge in the Introduction section of his decision (2006 NLTD 135):

[1] The Plaintiff is a retired registered nurse who resides in Fredericton, New Brunswick. James Karagianis and David Craig are psychiatrists ("Doctors"). The Health Care Corporation of

St. John's ("Corporation") is a hospital board which operates the Waterford Hospital ("Waterford") and the Health Sciences Centre.

[2] The Plaintiff's claims against the Doctors are that they negligently diagnosed her with a bipolar illness, negligently treated her for that illness and were guilty of medical malpractice.

[3] The Plaintiff's claims against the Corporation are in relation to her hospitalization at the Waterford in May and June of 1995 and consists of the negligence of its staff, breaches of the **Mental Health Act**, R.S.N.L. 1990, c. M-9 and the **Hospitals Act**, R.S.N.L. 1990, c. H-9, illegal confinement and the mishandling of the appeal of her involuntary certification.

[4] Karagianis was responsible for the care and treatment of the Plaintiff during her admission to the Waterford for the period of May 15, 1995, to June 19, 1995. He remained her psychiatrist (on an outpatient basis) until October of 1995. The Plaintiff alleges that Karagianis incorrectly made a diagnosis that she suffered from bipolar disease. The Plaintiff's allegations of negligence in relation to Karagianis (as set out in the Amended Statement of Claim) are as follows:

"(a) failing to properly review and substantiate the Plaintiff's medical charts, particularly in relation to a 1983 admission to the Health Sciences Centre, in the City of St. John's (the "Health Sciences Centre"), under the care of Dr. Jain, at which time the Plaintiff suffered the same symptoms and was diagnosed with and treated for a toxic drug reaction;

(b) failing to take proper account of the Plaintiff's statements of her condition and her warnings of her extreme sensitivity to drugs and to diagnose and treat her accordingly;

(c) failing to properly assess the Plaintiff and relying on improper information including collateral histories to make the diagnosis and give the treatment;

(d) relying on a second assessment for recertification which was improperly performed, completed, based upon erroneous information and without due consideration to the Plaintiff's statements and statements of immediate next-of-kin;

(e) failing to order tests and check (for) out other subjective voiced complaints (symptoms) which would reveal other potential physical and neurological conditions;

(f) failing to properly interpret and act upon the results of tests ordered, particularly where the said tests revealed abnormal results which would indicate the presence of other conditions; e.g. elevated Thyroid Stimulating Hormone;

(g) failing to consult with her previous treating physicians, in particular Dr. Jain, who had treated her for some eleven years prior to the admission, especially when his notes were illegible; and Dr. Wayne Button who was her family doctor for fourteen years;

(h) prescribing treatment, particularly drug therapy, of a kind and in amounts (which) that were dangerous and detrimental to the health of the Plaintiff, who was suffering from undiagnosed Fibromyalgia; and failing to take the Plaintiff off these drugs when serious side effects were relayed to him or ought to have been apparent to him".

[5] On November 15, 1995, the Plaintiff was admitted to the Health Sciences Centre and was placed under the care of Craig. She alleges that he incorrectly made a diagnosis of bipolar disorder. He treated her until she was discharged from the hospital on December 12, 1995, and he remained her psychiatrist (on an outpatient basis) until October of 1996. The

Plaintiff's allegations of negligence in relation to Craig (as set out in the Amended Statement of Claim) are as follows:

"In or about November 1995, the Plaintiff was admitted to the Health Sciences Centre, a hospital managed and controlled by the Third Defendant or its predecessor, and was placed under the care of the Second Defendant. The Second Defendant also negligently (diagnosed) assumed the Plaintiff as having Bipolar Disorder and negligently treated her for the same until up to about October 1996. The Plaintiff states that the details of this negligent diagnosis and treatment include:

(a) failing to properly review and substantiate the Plaintiff's medical history and to independently assess the Plaintiff's medical condition;

(b) failing to take proper account of the Plaintiff's statements about her condition and to order other tests and check for other symptoms which would reveal other potential physical and neurological conditions;

(c) failing to heed warnings of the Plaintiff's extreme sensitivity to drugs and to treat her accordingly;

(d) failing to consult with other treating physicians prior to and upon (making) assuming the diagnosis and rendering the treatment;

(e) prescribing treatment, particularly drug therapy, of a kind and in amounts which were dangerous and detrimental to the health of the Plaintiff and failure to take the Plaintiff off these drugs when serious side effects were relayed to him or ought to have been apparent to him;

(f) dangerously prescribing twice as much psychotropic medication as the patient needed as his opinion was that the plaintiff would only consent to taking half of what was prescribed;

(g) discharging the Plaintiff from the hospital while suffering from serious side effects of the drug therapy, nausea and vomiting and failing to properly remedy the side effects and assuming the vomiting was psychogenic;

(h) the Plaintiff further states that the Second Defendant breached the confidential relationship between the Plaintiff and the (Second) Third Defendant by relating to her employer his diagnosis of Bipolar Disorder, without her consent, as a result of which she has suffered damages".

[6] The Plaintiff also makes allegations of negligence against both Karagianis and Craig as follows:

"The Plaintiff repeats the foregoing and states that as a result of the negligent diagnosis and treatment by both the First Defendant and the Second Defendant, the underlying condition of the Plaintiff, that of Fibromyalgia as properly diagnosed in 1998, was not treated properly or at all, and the treatments prescribed by both the First Defendant and the Second Defendant were detrimental and dangerous to the health of the Plaintiff given the condition of Fibromyalgia. The Plaintiff states that on both occasions of admission, she was suffering, inter alia, from a toxic drug reaction which was not properly diagnosed or treated".

[7] In relation to her hospital admissions in 1995 at the Waterford, the Plaintiff alleges that the employees of the Corporation were negligent in the following manner:

(a) failing to properly assess and report relevant information to the First and Second Defendants or both of them;

(b) failing to properly, objectively and independently observe the Plaintiff, her condition and symptoms and report the same to the qualified physicians in charge;

(c) failing to properly obtain, review and substantiate collateral information;

(d) failing to provide proper and standard nursing care of personal hygiene required when a patient is too physically ill to care for herself;

(e) failing to act upon the filing of an Appeal for Release of the patient;

(f) failing to recognize physical symptoms of drug toxicity and not demonstrating as having the knowledge that drug toxicity can result from other underlying physical conditions; and

(g) failing to act in the professional nursing role as patient's advocate and bringing serious subjective complaints to the proper authorities or past the documenting phase. There were 63 entries on the progress notes of the plaintiff's concerns of drug toxicity.

[8] In relation to all three Defendants, the Plaintiff's Amended Statement of Claim also includes the following allegations:

"The Plaintiff states that the Third Defendant is vicariously liable of the negligence of its servants and/or agents and/or employees including but not limited to the negligence of the First and Second Defendants and employees responsible for observation, assessment, evaluation, obtaining and substantiating collateral information, reporting to the physicians in charge and care of the Plaintiff.

The Plaintiff repeats the foregoing and states that diagnosis (made) assumed and the treatment rendered was not in accordance with the standard of care, skill and attention required of medical practitioners with the education, qualifications, and experience of the First Defendant and the Second Defendant, and other health care employees and as a result thereof, the Plaintiff has suffered damages, including loss of income, loss of employment opportunity, loss of status and credibility, loss of professional mobility, damage to her professional and personal reputations, pain and suffering, extreme emotional stress and suffering and neurological damage.

The Plaintiff repeats the foregoing and states that the Third Defendant was negligent in failing to exercise due care in the employment, admitting privileges and/or retaining of staff so as to prevent negligent and substandard medical practices by the First and/or Second Defendant or other staff members when it knew or ought to have known that such failure would result in injury or damage to persons such as the Plaintiff.

In the alternative, the Plaintiff states that the Third Defendant was negligent in that it failed to fulfill its duty to provide or ensure adequate instruction, direction and supervision to its staff and in particular to the First and/or Second Defendant and persons to whom they and the Third Defendant entrusted the assumed diagnosis and treatment of the Plaintiff prior to the time or at the time that a diagnosis was made assumed and treatment was rendered to the Plaintiff.

The Plaintiff repeats the foregoing and states that if no individual Defendant is entirely responsible for the damages suffered by the Plaintiff, then the damages were caused by a combination of the negligence of all Defendants, and the Plaintiff pleads the Contributory Negligence Act, R.S.N. 1990, c. C-33, as amended.

In the alternative, the Plaintiff repeats the foregoing and says that the Defendants herein committed a breach of their contractual duties to the Plaintiff whereby the Plaintiff suffered damages including those set out herein.

In the further alternative, the Plaintiff repeats the foregoing and says that the Third Defendant through its employees, servants or agents committed a breach of the statutory obligations under the <u>Hospital Act</u>.

The Plaintiff repeats all of the foregoing and states that the actions of the Defendants, their agents, and/or employees amount to negligence towards the Plaintiff, trespass to the person of the Plaintiff and assault upon the Plaintiff, and in the case of the First and Third Defendants, illegal confinement <u>and treatment</u> contrary to the parameters mandated by the <u>Mental Health Act</u>.

<u>In this case, the plaintiff believes the causation of this assault stems from the denial of the plaintiff's right to an Appeal for Release to be objectively and independently heard by the Review Board. The errors written on the chart upon admission could have been identified, corrected and proper treatment, without injury, would have been the result</u>".

[9] The Plaintiff claims special damages, general damages, interest and costs.

[10] The trial commenced on January 11, 2005 and continued intermittently until May 25, 2006. There were a total of fifty-five trial days. There were thirty witnesses which included nine psychiatrists, a rheumatologist, a specialist in internal medicine, an emergency room physician, a neurologist, other physicians and various "lay" witnesses. In addition to the thirty witnesses, there were thousands of pages of documentary evidence which included hospital records, medical files, nursing notes and diagnostic reports in relation to the Plaintiff from June of 1983 until the commencement of this action.

[emphasis in the original]

[6] The principal witness for Ms. Day was Dr. Eleanor Stein. On motion of the Defendants, the Trial Judge ruled as follows (2005 NLTD 21):

[12] I find that [Dr. Stein's expert] Report is inadmissible as evidence in whole or in part because it:

(a) lacked independence and objectivity;

(b) contained pejorative and judgmental language;

(c) made legal interpretations and conclusions;

(d) was an instrument of advocacy and argument on behalf of the Plaintiff; and

(e) failed to confine itself to the appropriate area of expertise.

[13] Although the Report is inadmissible as evidence, this ruling does not preclude the Plaintiff from calling Dr. Stein to testify at this trial. ...

[7] In her testimony, Dr. Stein expressed the opinion that Dr. Karagianis and Dr. Craig misdiagnosed Ms. Day as having bipolar disease and that the medications they prescribed based on the misdiagnosis caused significant harm, due to toxic reactions to these drugs. Dr. Stein also expressed the view that Dr. Karagianis and the Health Care Corporation were complicit in "illegally detaining [Ms. Day] involuntarily at the Waterford [Hospital] during May and June of

1995" (Trial Judge's decision at para. 68). Dr. Stein concluded that Dr. Karagianis and Dr. Craig had been negligent professionally in their treatment of Ms. Day.

[8] Dr. John Martin, a rheumatologist, testified that Ms. Day had symptoms of fibromyalgia and chronic fatigue syndrome. (Dr. Stein testified to similar effect.)

[9] Expert witnesses for the Defendants were Dr. Keith Pearce and Dr. Michael Teehan. Their views were that the psychiatrists who diagnosed Ms. Day with bipolar disease, including Dr. Karagianis and Dr. Craig, made an accurate diagnosis.

[10] In his decision (at paras. 73-74), the Trial Judge set out excerpts from Dr. Pearce's expert report. They read, in part:

> At least nine different psychiatrists (Doctors Pantel, Jain, Karagianis, Pratt, Zielonka, Craig, Frecker, Doucet and Singh) have recorded their diagnostic impression of Ms. Day in the records provided to me. All are in agreement that bipolar disorder was the appropriate diagnosis. <u>I do not find a single contrary opinion from a psychiatrist or physician or in any of the nurses' progress notes.</u> ...

> The records maintained by Doctors Karagianis and Craig and varied hospital staff, which I have reviewed in detail, <u>in my opinion reflect the highest standards of psychiatric care, including diagnosis, the most up-to-date treatment for the bipolar disorder diagnosis they made, and a level of human concern for their patient's welfare which was of the highest ethical standard.</u> Since Doctors Karagianis and Craig treated Ms. Day, several other psychiatrists have attempted to treat her with a similar lack of lasting success.

[11] In para. 78 of his decision, the Trial Judge set out the following excerpt from Dr. Teehan's expert report:

> Having surveyed the various hospital charts, the detailed clinical recordings by the physicians, nurses and other clinicians involved in the care of this patient, <u>I have no doubt that she has quite a classic Bipolar Mood Disorder. Not only are the clinical features described typical of episodes of mania, depression, and mixed episodes, but the pattern of her illness is also classical, with recurrent episodes, apparently some of them provoked by stressful events in her life.</u> Unfortunately, the patient has rejected the diagnostic conclusions of several different psychiatrists, who have all agreed that the diagnosis is correct. As mentioned earlier in this report, there has been consideration of other potential causes of the changes in behavior and thinking that she has experienced. The patient has proposed at various times that her illness was caused by antihistamine medication, and laterally by fibromyalgia. To accept the proposition that her illness was caused by antihistamine medication one would have to accept that several different compounds produced identical clinical features to mania; identical episodes separated by 12 to 14 years, <u>and that as a separate issue she developed periods of depression.</u>

> Ranked against this possibility is the opinion of a number of unrelated Specialists in Psychiatric Medicine, who separately and without collusion identified the presence of features of Bipolar Mood Disorder, on separate occasions, and who showed evidence in the clinical notes of their work that they carefully evaluated her history and mental state and relevant investigations in making that diagnosis.

> In evaluating the lengthy clinical records in all of the written material pertinent to this case, I am also faced with the inescapable conclusion that this patient has very clear cut history and

symptom picture of Bipolar Mood Disorder. Tragically, and possibly as part of the manifestation of that illness, she is unable to recognize or accept this diagnosis.

...

I would therefore have to say that I find no evidence that either Dr. Karagianis or Dr. Craig deviated from the expected standard of care for Psychiatrists managing a case of this kind, and that they have followed well accepted clinical paths of care for this challenging case.

[12] The Trial Judge went on to weigh the evidence and state his conclusions, as follows:

[80] The testimony of Dr. Stein is diametrically opposed to the testimony of Drs. Pearce and Teehan. In determining the facts of this case, I find that the testimony of Dr. Stein is the most problematic.

[81] As with the nature and quality of her Report (which was ruled inadmissible), I also have concerns about Dr. Stein's qualifications and experience in relation to the medical issues presented at this trial. In addition (throughout her evidence), it appeared that she was testifying more as an advocate for the Plaintiff rather than an independent and objective expert witness.

[82] Although Dr. Stein was qualified to practice as a psychiatrist in 1992 she only practiced adult psychiatry for a period of one month. She had little or no experience in acute adult psychiatry. She never certified an adult patient for involuntary admission to hospital and she never cared for an acutely ill adult patient from the time of admission through to the time of discharge from the hospital. Her testimony in relation to involuntary certifications was based on her limited experience when she practiced in the provinces of Ontario and Alberta. Those provinces have legislative criteria for determining certifications that are significantly different from those in this province.

[83] Since 2001, Dr. Stein has been involved exclusively in the treatment of patients with Fibromyalgia and Chronic Fatigue Syndrome.

[84] I have considered the evidence of the three expert witnesses in relation to their qualifications, their experience and the evidence presented at this trial in relation to the issue of bipolar disorder. On the basis of the examination of the record, I accept the testimony of Drs. Pearce and Teehan and reject the evidence of Dr. Stein. As a result, I do not find that Karagianis and Craig breached their duty of care in relation to the Plaintiff.

...

[86] Accordingly, on the basis of the evidence and especially the evidence of Drs. Pearce and Teehan, I find that the Plaintiff failed to prove any of the allegations of negligence set out in her Amended Statement of Claim in relation to Karagianis and Craig.

[13] The Trial Judge went on to dismiss the action against the Health Care Corporation of St. John's, and also found that Ms. Day had not been illegally confined (as she had alleged) in the Waterford Hospital.

[14] In his summary, the Trial Judge stated:

[99] The Plaintiff has had psychiatric problems since 1983. Between June of 1983 and January of 1998 she was admitted to hospital on seven occasions for manic and/or psychotic episodes. She was involuntarily "certified" under the Mental Health Act, supra, on four of those hospitalizations.

[100] The Plaintiff's psychiatrists diagnosed her with bipolar disorder/affective disorder on six of those hospital admissions. Between March of 1997 and October of 2000 the Plaintiff was treated

on an outpatient basis by a psychiatrist who saw the Plaintiff on twenty-six occasions during that period of time. He observed both mania and depression over that three-year period and treated her for bipolar disorder.

[101] The Plaintiff has failed to prove (on a balance of probabilities) any of the allegations of negligence as set out in the Amended Statement of Claim. The Plaintiff's claims against the Defendants are dismissed.

[102] In the circumstances of this case, I do not find that it is appropriate to award costs.

ISSUES

[15] Ms. Day appeals the Trial Division decision. In her factum she stated the issues as follows:

(1) The learned judge erred in law and in fact in that he failed to consider whether the Defendants negligently omitted to include medication effects in their working or final diagnosis.

(2) The learned judge erred in law in that he failed to make a provisional assessment of damages.

(3) The learned judge erred in law in that he failed to provide reasons for his decision sufficient to permit meaningful appellate review. There was evidence, to support my case, presented to him that was not mentioned on (sic) his decision, so this has prejudiced me from initiating a proper appeal.

ANALYSIS

[16] At the outset of the hearing, the Chief Justice outlined briefly that it is not the role of the Court of Appeal to re-hear the case and substitute its own decision for that of the Trial Judge. Rather, where an appellant takes the view that the Trial Judge has made an error of fact, it is necessary for the appellant to show that there had been a "palpable and over-riding" error. Where an appellant takes the view that the Trial Judge has made an error of law, it is necessary for the appellant to show how this is so. The Chief Justice made clear that the members of the panel had read Ms. Day's written submissions. Finally, the Chief Justice indicated that while Ms. Day should address herself to the issues on appeal, leeway would be accorded to her, as she was representing herself.

(1) "The Defendants negligently omitted to include medication effects in their working or final diagnosis."

[17] In her submissions, Ms. Day urged on the Court the view that her problems had arisen from the toxic effects of medication and not from having bipolar disorder. She referred to the support for this view in the testimony of Dr. Stein. She also referred to her own testimony as to her improvement since she ended treatment (especially medication) by Dr. Karagianis and Dr. Craig.

[18] Mr. Day submitted to the Court that the Trial Judge had erred in not accepting the expert opinion of Dr. Stein and, instead, accepting those of Dr. Pearce and Dr. Teehan.

[19] Ms. Day, in particular, attacked the credibility of Dr. Pearce, that his expert opinion had not been accepted in the following cases in Alberta.

[20] In **Baillie v. Crown Life Insurance Company** (1998), 59 Alta. L.R. (3d) 45 (Q.B.) at para 33 et seq., Clarke J. took issue with Dr. Pearce's view that chronic fatigue syndrome "does not exist".

[21] In **Phillips & Phillips v. Rost** (1996), 40 Alta. L.R. (3d) 246 (Q.B.) at para. 152, Moreau J. rejected Dr. Pearce's opinion that the plaintiff was not "forthright" and Dr. Pearce's attribution of the plaintiff's "on-going problems to hysteria, a desire for secondary gain, and malingering".

[22] In **Reynolds v. Pohynayko et al.** (1997), 202 A.R. 1 (Q.B.) at para. 53, Bielby J. wrote, "Plaintiff's counsel launched an effective attack on the objectivity of Dr. Pearce, a psychiatrist retained by the Defence to test and interview [the plaintiff]."

[23] In **Stevens v. Okrainec** (1997), 210 A.R. 161 (Q.B.) at para. 254, Nash J. stated, "I do not accept the opinion of Dr. Pearce that [the plaintiff] has pain of a non-organic origin that was caused by her personality problem …".

[24] Do these cases undermine Dr. Pearce's credibility? They certainly cast a shadow over it. I will assume, without so deciding, that the Trial Judge should not have been persuaded by Dr. Pearce's expert evidence. That does not undermine Dr. Teehan's expert evidence. Ms. Day is left in the position that the Trial Judge found Dr. Teehan's evidence to be credible and he found Dr. Stein's evidence not to be so. The Trial Judge set out a reasoned basis for his acceptance of Dr. Teehan's expert opinion and for not accepting that of Dr. Stein.

[25] Notwithstanding Ms. Day's submissions that we should reverse the Trial Judge as regards his exclusion of Dr. Stein's expert report and as regards his finding that Dr. Stein's testimony was not credible, Ms. Day failed to provide any sufficient basis for this Court to reverse the Trial Judge on these matters. Accordingly, on the first issue, Ms. Day's appeal fails.

(2) "Fail[ure] to make a provisional assessment of damages."

[26] The Trial Judge did not err by failing to make an assessment of damages when he dismissed Ms. Day's claim. On the second issue, Ms. Day's appeal fails.

(3) "Fail[ure] to provide reasons for his decision sufficient to permit meaningful appellate review."

[27] The essence of Ms. Day's submission is that the Trial Judge did not deal in his decision with key aspects of the evidence submitted on her behalf. While little of Dr. Stein's testimony is specifically dealt with in the Trial Judge's decision, the essential thrust of her evidence is. As well, the Trial Judge provided a reasoned basis for finding that her testimony was not credible. He was not required to do more. See **R. v. Sheppard**, [2002] 1 S.C.R. 869. Accordingly, on the third issue, Ms. Day's appeal fails.

THE CROSS APPEAL

[28] The Defendants have cross-appealed the Trial Judge's decision not to award costs. To recall, he wrote at para. 102:

In the circumstances of this case, I do not find that it is appropriate to award costs.

[29] The Defendants concede that an order as to costs is discretionary, but submit the foregoing is contrary to principle and should be reversed.

[30] The Trial Judge did not set out a rationale for departing from the ordinary rule that costs are awarded to a successful party. However, one can infer that the circumstances the Trial Judge had in mind relate to Ms. Day's situation, especially that she has devoted so much effort and (as she

told the Court) all her financial resources to this case. Is that a valid basis for denying the Defendants their costs?

[31] Mark Orkin, in **The Law of Costs**, 2d ed., loose leaf (Aurora, Ont.: Canada Law Book, 2007) vol. 1 at pp. 2-78 to 2-80, lists several exceptions to the rule that a successful party is entitled to their costs, including misconduct of the parties, miscarriage in the procedure, and oppressive and vexatious conduct of proceedings. Orkin then states:

> The fact that the imposition of costs would cause financial hardship is not sufficient to displace the ordinary rule that costs should follow the event. As has been said, to make a determination of costs based on the respective financial positions of the parties would undermine the logic and purpose of the rules. Neither the needy circumstances of an unsuccessful plaintiff nor the defendant's financial ability to absorb the costs are relevant considerations. ...

[32] Orkin indicates that it is appropriate for a judge to take into account the tragic circumstances of a case when exercising discretion as to costs; for example, where the unsuccessful party needs medical help or where imposing costs would impose a severe hardship on the unsuccessful party (p. 2-80). However, at p. 2-81, Orkin states:

> ...the fact that a trial judge feels sympathy for an unsuccessful plaintiff is not an appropriate reason for departing from the general rule that costs follow the event, although a variation from the rule may be appropriate to avoid injustice to a party.

[33] **Murphy v. Myers** (1995), 127 Nfld. & P.E.I.R. 86 (NLTD) [affirmed on appeal at (1997), 157 Nfld. & P.E.I.R. 323 (NLCA), the costs issue not being at issue in the appeal] dealt with a claim for damages for personal injuries allegedly arising out of a motor vehicle collision. Justice Green (as he was then) dismissed the claim. With respect to costs, the plaintiff argued that the imposition of costs would cause him severe financial hardship. At para. 10, Green J. stated:

> Even accepting that the plaintiff acted in good faith in bringing his proceeding and that the claim is not a frivolous one I am not satisfied that it is appropriate in this case to deprive the defendant of the costs to which, as the successful party, he would normally be entitled solely because the defendant is better able to bear them.

The defendant was awarded costs on a party and party basis.

[34] **Brown v. Black Top Cabs Ltd.** (1997), 97 B.C.A.C. 59 (BCCA) involved an action for negligence arising from a motor vehicle accident. The apportionment of liability and its variation on appeal gave rise to a complex result. However, the Court set out a simple rule in the context of hard circumstances. The plaintiff was a single mother, supported by social assistance, unemployed and probably unemployable as a result of the accident. She argued that costs should be not awarded against her in light of her circumstances. At paras. 16-17, the Court stated:

> We are of the view that a discretionary award of costs of the sort sought here seeks to elevate needy circumstances to a principle of law. ...
>
> ... The considerations to be borne in mind should be those that arise from the nature and conduct of the litigation. ...

In my view, the Court acted in accordance with established authority in the foregoing passage. See also **Wilson v. INA Insurance Co. of Canada** (1998), 112 B.C.A.C. 208 at para. 25.

[35] There is nothing to suggest that the Defendants in this case conducted themselves in the course of litigation so as to warrant being denied their costs; nor is the case of a nature (e.g. a test case on a matter of public policy) that would constitute a basis for denying a successful party its costs. Accordingly, the difficult circumstances of the Plaintiff are not a basis to deny the Defendants their costs.

[36] <u>I would grant the cross-appeal and award the Defendants their costs on a party and party basis in the Trial Division</u>. Having been successful before this Court (both on the appeal and cross-appeal), the Defendants will have their costs here, also on a party and party basis.

_____

M. Rowe, J.A.

I Concur: _____

C. K. Wells, C.J.N.L.

I Concur: _____

B. G. Welsh, J.A.

    I know that the outcome of this lawsuit in Newfoundland and Labrador is wrong and unjust. I have learned through this unprecedented experience of spending almost sixty days in a courtroom seeking justice that what is legal is not necessarily what is just. In my quest for justice, I still have faith in the justice system of this country and believe someday that justice will prevail for the injured people of medical malpractice, especially psychiatry as new precedents are created and set.

I am appealing this case to the Supreme Court of Canada on the following grounds:

- Ground 1 The trial judge erred in his decision not to allow my medical expert, Dr. Stein's report to be resubmitted, with the pejorative language removed.

- Ground 2 The trial judge erred in his subjective opinion that Dr. Stein was not qualified and lacked experience to diagnose and treat the condition of bipolar disorder, which influenced his decision not to rule in my favor.

- Ground 3. The trial judge erred in his subjective opinion that Dr. Stein was biased, without giving objective evidence why that was his belief.

- Ground 4. The judges in Newfoundland and Labrador erred in accepting "only" the evidence presented by two psychiatrists for the defense, and not considering all the other medical evidence presented by other specialists, family and friends when making their decision.

- Ground 5. The Trial judge erred in accepting Dr. Pearce's testimony because of his shady past record in Alberta.

- Ground 6. The trial judge misapprehended evidence to come to the wrong conclusion as is written in the summary, par 99 and 100.

- Ground 7. The Appeal judges erred in their assumption that the trial judge did not award costs due to financial circumstances and erroneously overruled the trial judge's decision not to award costs.

- Ground 8. The trial judge was witness to privileged information that confirmed Physicians lied under oath. Armed with this knowledge, the circumstances were present not to award costs.

The application was rejected by the Supreme Court of Canada so I had reached a glass ceiling. I still felt confident about my quest for justice and satisfied that I had taken it to the limit, but I believe the justice system of Canada had also failed me.

The defense sent me a bill for $180,000 shortly after they received news that my application for Leave to Appeal was denied. Another bill came within a month or so with over $3,000 of interest added to the $180,000, so I went immediately to a financial agency to file bankruptcy.

Within less than a year I had fulfilled all my requirements set down by the Bankruptcy Act and was released from that strong hold, for which I had prepared myself. Because I owed no other money to the banks or credit card companies, I had my credit rating restored and was allowed to hold a credit card.

In bankruptcy, I was able to keep my home that I adore, and could continue living in my warm and loving home, where my friends and family are always welcomed with a friendly smile and lots of food and drink.

The only assets I had worthwhile were my car and my hot tub. I bought back my car and was allowed to keep my hot tub as it was a medical necessity. By reducing my expenses, and taking in a boarder, I managed to pay off that bill and became free and clear of that $180,000 debt that I would never have been able to pay otherwise.

When all the expenses were disbursed by the bankruptcy agency, there were $3500 left for the CMPA, but they did not accept the money so it was reverted back to me just before Christmas, 2010. This windfall encouraged me to write this book as I have some extra money now, and I could not think of a better use for it than that.

I have survived three years 1995-1998 in a psychiatric trap. I have lived over ten years, (1998-2008) without any psychiatric interventions or toxic medications, carrying on with my new life of advocacy, education and good will; fighting a court battle to obtain redress and financial compensation for medical negligence causing harm. In 2009 I survived bankruptcy.

Now in 2011, I have relived the whole experience while writing this book, which will be the ultimate cleanse and my legacy. I will bring closure to sixteen years of hardship that led to incredible personal growth, which has made me into the person I am today, with the will and desire to persevere and evolve. I believe I have finally won-won, living a life that is full and blessed with rewarding accomplishments and success.

Yes, I am well aware that I have made enemies, especially within the medical establishment, but so did Gandhi and Jesus Christ himself. All I can say about that is "We must forgive each other and ourselves," leaving the record-keeping up to God, the ultimate and final judge." "As you have sown, so shall you reap."

The incorrect diagnosis of Bipolar Disorder had its beginnings May, 1995 when, of all things, my second ex-husband was interviewed by the Clinical Clerk, who had very little psychiatric experience, and thus, I am sure now that he gave damaging incorrect information to her to support the diagnosis of Bipolar Disorder which had never been established before that time. I have no doubt in my mind that this is what happened and why the error in diagnosis was made, as he has committed the same tort fifteen years later on Thanksgiving Weekend 2011.

Since the first edition of this book was published in September, 2011, my second ex-husband has again come forward to slander my good name and declare to strangers that indeed I am mentally ill with Bipolar Disorder, not credible and he even tried to convince others not to let me speak in public about my book <u>Judging Judi</u>. Furthermore, the publication of the first edition of <u>Judging Judi</u> was stopped, so I sought legal advice and have made alterations in content so hopefully that this book can once again be published.

After all this time, this man is still trying to damage my credibility and my life, but he has not succeeded. I have indeed moved on and I know people believe in me as I have always believed in myself.

I definitely know that I am in a win-win situation now as I have realized all the answers and this time the questions will remain the same. On this special Remembrance Day in Canada 11/11/11, I submit this manuscript for publication of the second edition. Lest we forget.

Injustice anywhere is a threat to justice everywhere.

It is highly convenient to believe in the infinite mercy of God, when you feel the need of mercy, but remember also His infinite justice.

Martin Luther King Jr.

Newfoundland and Labrador Court of Appeal

Epilogue & Summary

Psychiatrists are considered experts in a field of medicine where diagnosis of psychiatric illness is based mainly on subjectivity with no laboratory analysis or proof. Still, not many people feel they can question these experts or bring them to task for their expert opinions. "After all it is 'only' their opinion." The judge will say. A psychiatric opinion can be very damaging.

But it appears psychiatrists cannot be wrong. They freely throw around their damning expert opinions with no fear of being sued for defamation of character or other insulting remarks. They are protected by their peers, their insurance company (CMPA) and the courts.

According to Dr. David Craig, psychiatrists are wrong in 25%-30% of their diagnoses. Over one quarter of the people they manage are incorrectly diagnosed and are treated with dangerous drugs that change the brain chemistry, which may result in serious repercussions.

Psychiatry unto itself is an inexact science of medicine, which has been mostly practiced by men until these past few decades. Now in the twenty first century I believe this medical practice needs to be controlled. The psychiatrists have to be held accountable when serious injury results from prescribing dangerous psychiatric medications, especially when the FDA rules and regulations are not followed and irreparable damage results.

Many times, drugs are prescribed by physicians "off label" (not for what the drug was intended) or without recommendation by Compendium of Pharmaceutical Specialties, (CPS). Controlled testing 'must' be carried out by the pharmaceutical industry on the elderly female population or the very young, before the approval can be given for the drug to enter the market and be freely prescribed to those individuals. Even general practitioners can prescribe those dangerous psychotropic drugs, and in most cases, women are the main recipients as usually they live longer than men, and according to statistics, a substantially higher number of women suffer from the chronic illnesses of arthritis and fm/cfs/mcs.

Most drugs, including psychiatric drugs, have only a six week trial time before being approved. Then they are prescribed for long periods, with serious consequences, resulting in permanent neurological and other physical damage. People must be monitored more closely for side effects.

Drug trials for the atypical antipsychotics, especially ZYPREXA, were done in Newfoundland and Labrador and in other northern regions of Canada, with serious consequences to the patients, which resulted in out of court settlements by the drug company Eli Lilly.[1] The fines were minor in comparison with the damage done to the people who were prescribed them.

The incidence of drug interactions and overdoses is alarming and taking peoples' lives, let alone the unnecessary hospital admissions and disabilities that result.

The elderly, mostly women living in nursing homes are prescribed those dangerous drugs to relax them when they become upset and agitated, and children are being given psychotropic drugs to keep them quiet in school, without knowing what consequences they are having on their developing and aging brains. There has and must be a better way of treating people without the potential risk of damaging them.

[2]The medical experts of human behavior, who were mainly males, have the power to control people's lives, especially women's, within the medical practice of Psychiatry. These physicians have

---

[1] Zyprexa Class Action, Canada Wide Settlement, May 5, 2010, (schedule F)
[2] **Outrageous Practices**, the alarming truth about how Medicine Mistreats Women. Leslie Laurence and Beth Weinhouse.

the empowerment to define mental illness, even though "men" have been mainly studied, and created normative theories concerning all aspects of life. Men's and women's bodies are as different as chalk and cheese, and react differently to chemicals and the amounts given.

If there is no scientific evidence that can prove that a patient has a mental illness, eg. bipolar disorder or chemical imbalance, and the patient is living productively with no threat to anyone, why can the patient not be given the benefit of the doubt? Why would forced drug therapy be allowed to happen?

Is a mental status "abnormal" just because a person does not agree with a psychiatrist's diagnosis, in a practice of medicine where 25% to 30% of their diagnoses are incorrect?

I was never certifiable with two signatures again after May 30th, 1995 for having 'no insight' into or understanding of my supposed mental illness. Why was I recertified then? I had a normal mental status otherwise.

[1]If you place an individual with a normal mental status among the mentally ill in confinement, supposedly experts cannot tell the difference between who is mentally ill and who is not. If this is the case, why is a patient so removed from the decision of treatment and not listened to when he/she is admitted to a psychiatric unit? Why can a patient not be present at rounds when his/her behavior is being discussed? Why can the patient not have the chance or right to explain incidents that have happened and may be recorded as inaccuracies? Why can a patient not be present when these incidents are discussed during medical 'rounds"? What is wrong with laughing out loud when you hear a joke or shedding a few tears when you read a sad poem? Emotions are normal; but flat emotions or being reduced to a zombie with medication is not normal, whatever normal is. Dr. Doucet told me, during one of our consults "In psychiatry, 'normal' was considered a setting on a dryer." Then why do psychiatrists give out labels so easily?

When I was admitted to a psychiatric unit because of adverse reactions to drugs, anxiety due to extreme stress or other emotional battering, I was treated like an invisible uncommunicative person, who could not speak for myself or have the chance to explain how I was feeling, and the misunderstandings kept happening.

What gave a psychiatrist the right to yell at me and point his finger at me without even consulting me about what was wrong? What gave the psychiatrist the right to write inaccuracies in a chart to suit the diagnosis he believed I had because of previous recorded inaccuracies?

I believe also that the four models of therapy, which are the medical model, the psychological model, the cognitive/ behavioral model and the social model should be integrated and neither one of them should stand alone. The problem now is the medical model is standing alone in our health care system.

When people are being treated holistically, (mind, body and spirit), one must consider the four models as above. I believe people must be individually assessed by a professional and treated according to their medical, personal and social history. Patients could be referred to social workers, counselors, occupational therapists, massage therapists and psychologists and treated accordingly, with a much better outcome, and at a reduced cost.

If psychiatrists do not have the time to treat people from head to toe, then other professionals must and can meet this challenge. These other professional services must be made affordable to everyone if we are going to treat people appropriately and effectively, and save money on our socialized health care system that presently concentrates on illness, instead of wellness.

Drug therapy is a palliative treatment which causes many people in society to never recover from an emotional stressor and to become addicted. These drugs replace the body's own ability to secrete these neurotransmitters and hormones, thus people are completely dependent on chemicals to live.

---

[1] The Famous Rosenhan Experiment into the validity of psychiatric diagnosis conducted by psychologist David Rosenhan in 1973.

These people are not really living any more, only existing in a fog of artificial emotions that will eventually lead to their untimely deaths. However, there are times I know drugs are needed, and thanks to drugs, it enables some people to live longer and have healthier lives, but patients have to be aware of and monitored for side effects of those drugs before damage results.

If psychotropic drugs are used frivolously and for long periods of time, some people will be doomed to years of drug dependency and ill health. They will have the battle of their lives trying to wean themselves off the chemicals and back to the reality of living life normally, as it is meant to be lived.

What is a normal life anyway? It is what the dominant group states it is. It is the version of the world that has been sanctioned as reality. It has no room for turning to God in life's unbearable moments, or losing control and breaking down under the strain of an intolerable existence, or when your whole life and reason to be are being threatened by external negative forces resulting from a severe shock, imprisonment or being controlled. And then have the worst happen "to be stigmatized and labeled with a mental illness forever".

These out-cries of emotions are reactions to severe stress from an intolerable lived experience and are nature's way of protecting people to survive harsh and unbearable environments and situations, so they can have the ability to heal and move on. But they will do neither as long as their brain and behavior is being controlled with drugs.

These psychotropic drugs are being used as a powerful modality of treatment for social control, without taking into consideration other social issues that may arise such as financial survival, prejudice, oppression and inequality, which create stress, conflict, difficulties and unhappiness. These symptoms are viewed by the psychiatrists as more symptoms of a mental illness. Prescriptions are given to patients that will be viewed as a solution to their misery, but at what cost?

The cost was great to me and to society as in the meantime I was reduced to living without an income, forced to live in one room of a boarding home because of being too "ill" from being drugged to physically care for myself any more. It took every ounce of energy to live and breathe, while trying to maintain or regain my autonomy, identity, and health. Psychiatry took away everything from me, and gave me back a label and a prescription to treat that label and the rest was up to me to recover from the damages of the medical model of psychiatric care.

View of the St. John river from my balcony in Fredericton

The New International Consensus Criteria for ME

There has been a recent publication of new criteria for myalgic encephalomyelitis (ME). The new criteria designate ME as the appropriate biological name for the illness. The name Chronic Fatigue syndrome (CFS) has been put to rest, as it was confusing and too much focus was given to the symptom of fatigue, which is only one of many.

In this book I have mentioned the 2003 Canadian Consensus Criteria that were used as a starting point to make a diagnosis, but recently significant changes have been made. Instead of the cardinal symptom of the newly defined ME being "pathological fatigue" or "post exertional malaise," it will be now named "post-exertional neuro-immune exhaustion." There is no longer a six month waiting period required before diagnosis can be established.

The biological focus for the new definition of ME is clear, consistent, and unequivocal. The ultimate value of the new definitions will be determined by the researchers, clinicians, and policy makers. There are no ready answers about the impact of the re-defined ME on important issues such as illness credibility, physician willingness to diagnose and treat ME, and disability determinations, but I am positive that this change is best for all. The National ME/FM Action Network was ahead of its time, when they chose its name.

The kind of man who always thinks he is right, that his opinions, his pronouncements are the final word, when once exposed shows nothing there. But a wise man has much to learn without loss of dignity./Sophocles

My Garden, Summer 2011

Carruthers BM, van de Sande MI, (co-editors), De Meirleir KL, Klimas NG, Broderick G, Mitchell T, Staines D, et al. Myalgic Encephalomyelitis: International Consensus Criteria. J Intern Med 2011 Jul20. Doi: 10.1111/j.1365-2796.2011.02428x

What do I want from my physician or health Care Worker?

- *I want support help and advice, not orders and demands.*
- *I want to leave the doctor's office or hospital feeling better about myself than when I entered it.*

- *I want understanding and empathy as to my behavior, thoughts and feelings.*
- *I want to be accepted and respected as a human being, not labeled with an illness.*
- *I want a good listener who is empathetic and appears genuinely concerned about me and my wellbeing.*
- *I want someone who takes all the time it needs to listen to me so that both of us can determine what is wrong. Nobody is genius enough to understand my problem whether it is physical or psychological in five minutes.*
- *I want someone who prescribes medications, including tranquilizers, antidepressants, mood stabilizers and antibiotics, after the proper tests are done; and as the last resort.*
- *I want to be informed unconditionally, and be actively involved in decision making regarding any required therapy.*
- *I want a physician or health care worker who is able to admit when he/she has made an error in diagnosis and/or treatment, even if it resulted in injury.*
- *I want to have a relationship with my physician or health care worker that if a mistake does occur that litigation would be the last thing I would consider.*

A doctor who cannot take a good history and a patient who cannot give one are in danger of giving and receiving bad treatment. /Author unknown

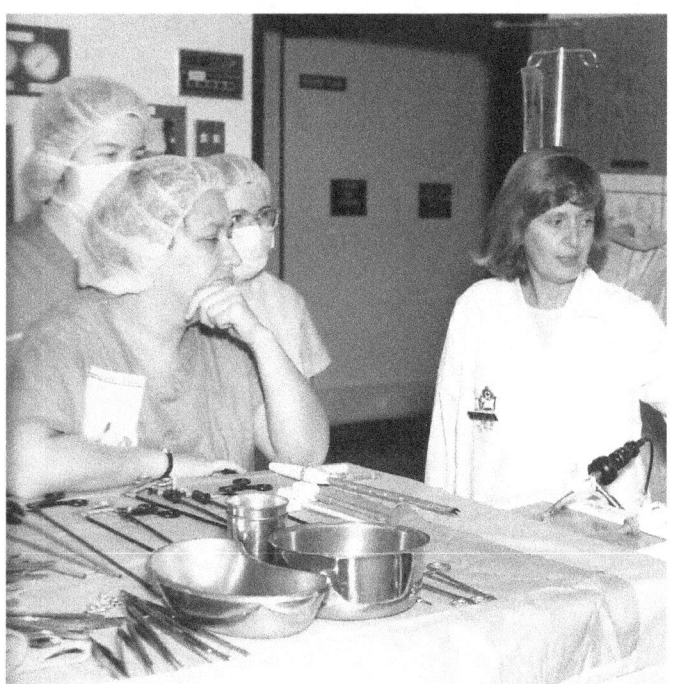

This photo was taken by the Telegram when we hosted Open House in the Operating Room, at the Health Science Centre in the early nineties. I was demonstrating the instrumentation involved in laparoscopic surgery.

About the Author

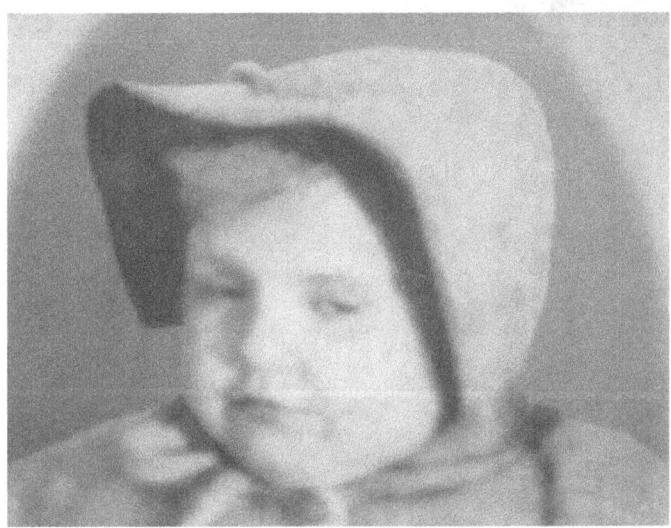

Judith Day lives in Fredericton, New Brunswick.

Mother of two adult children who are married and independent. She has three grandchildren.

She spent most of her nursing career in the operating room and served on the executive of professional associations in nursing and other quality assurance committees, developing QA programs, conducting nursing audits and filing reports.

After being medically retired in 1998, she facilitated a Fibromyalgia, Chronic fatigue Syndrome Self Help Group of Newfoundland and Labrador for five years 1999 to 2004 before moving to Fredericton and presently is Director for Atlantic Provinces for National ME/FM Action Network.

She presented herself for candidacy with the Canadian Alliance in 2000, but with a history of mental illness was not considered credible, so her aspirations for a political life were short lived.

To prove her mental stability, she went public again and ran as deputy mayor and received over 5,000 votes from the people of St. John's, which restored her confidence that she was indeed credible, but retired again from politics, instead of running for Councilor at Large as it would be too physically demanding due to her chronic condition and fluctuating limited energy.

While living with and managing her chronic illness of FM/CFS/MCS, she keeps herself busy with volunteer work and is an advocate; educator and activist. She remains involved in public matters, but paces herself constantly to remain healthy. She writes and speaks out on health, welfare and other social issues that affect us all.

Judi is able to participate in sports like skiing, bicycling, swimming, kayaking and of course she enjoys gardening, dancing and walking, all in moderation, with many rest periods.

Although the legal and ethical definitions of right are the antithesis of each other. Most writers use them as synonyms. They confuse power with goodness and mistake law for justice./ Charles T Sprading

Judi

You had the courage to put yourself forward for service in the public domain. That in its self is a victory regardless of the bottom line in the vote.

You are a valuable and worthwhile person. Thank You for continuing to give to us all.

Bill

www.ingramcontent.com/pod-product-compliance
Lightning Source LLC
Chambersburg PA
CBHW080905170526
45158CB00008B/2001